Carbon Dioxide Capture

Innovations in Advanced Material Interface

Santosh K. Tiwari, PhD

Centre for New Materials and Surface Engineering,
Nitte (deemed to be university), Karnataka

Irfan Ayoub, PhD

Department of Metallurgical Engineering and Materials Science,
Indian Institute of Technology Bombay, Mumbai, 400076, India

notionpress.com

INDIA · SINGAPORE · MALAYSIA

ISBN
Paperback 979-8-90319-706-4
Hardcase 979-8-90540-641-6

TABLE OF CONTENTS

FOREWORD

It is a matter of great pride and satisfaction to present this scholarly volume titled ***"Carbon Dioxide Capture: Innovations in Advanced Material Interfaces",*** which addresses one of the most urgent global challenges-mitigating atmospheric carbon dioxide and its impact on climate systems. The transition toward sustainable energy and environmental stewardship requires scientifically robust and technologically scalable solutions, and advanced materials play a central role in this transformation.

This book provides a comprehensive and timely exploration of material interfaces governing the adsorption, absorption, and catalytic conversion of carbon dioxide. By integrating principles of surface engineering, surface chemistry, and nanostructured systems, the authors highlight innovative pathways for enhancing the efficiency, selectivity, and long-term stability of carbon capture technologies. The emphasis on interface-driven phenomena is particularly relevant for next-generation carbon management strategies. Such contributions are essential not only for advancing academic research but also for guiding industrial implementation and informing policy frameworks .

I commend Prof. Tiwari and Dr. Irfan for this rigorous and impactful work, which will serve as a valuable reference for researchers, educators, and practitioners engaged in sustainable materials and climate solutions. I extend my sincere congratulations to the authors and wish them continued success in making many more meaningful contributions to the academic world in the years to come.

Prof. Dr. M.S. Moodithaya
Vice Chancellor
Nitte (deemed to be university), Karnataka

PREFACE

It gives me pleasure to note that Dr. Santosh K. Tiwari of Centre For New Materials & Surface Engineering, NMAMIT, Nitte University, Karnataka, India, has written a book titled *Carbon Dioxide Capture: Innovations in Advanced Material Interfaces*, which is an outcome of his Government of India funded SPARC (Scheme for Promotion of Academic & Research Collaboration) project titled (*Layered Nanocomposite Composed of Graphene Carbon Quantum Dots and Mxene Coated with Calcium Oxide Nanoparticles for Large-Scale Carbon Dioxide Capture and Translation into Nano-Fertiliser*), which was jointly with (University of Nottingham, UK). The SPARC Program is aimed at fostering high-quality, outcome-oriented research through international academic collaboration, which, most importantly, involves student participation and mobility to the laboratories of the partner International Institutes.

Coming to the theme of the book, *Carbon Dioxide Capture* is not only a scientific challenge but also a major priority for nations striving towards sustainable development and climate resilience. The distinguishing strength of this book lies in its focus on the role of advanced material interfaces as the central element governing CO_2 capture performance. This perspective reflects the maturity of the field, where incremental improvements in bulk materials are no longer sufficient, and transformative advances are increasingly driven by interface-level design and control. By addressing how chemical functionality, electronic structure, porosity, and transport phenomena reconverge at interfaces, the authors present a scientifically rigorous and forward-looking narrative. Indeed, I am happy to note that this book emphasises the interdisciplinary nature of the topic, seamlessly merging fundamental science with real-world applications, focusing on the integration of materials chemistry, surface physics, electrochemistry, and chemical engineering, with a strong focus on energy efficiency, operational stability, and scalability. I particularly appreciate the attention given to emerging hybrid and multifunctional systems, including electrochemical modulation and selective transport. I am sure that for students and early-career researchers, this book will serve as a great learning resource. Even for established researchers and policymakers, the book offers a clear demonstration of how targeted investment in materials and interface science can contribute meaningfully to climate mitigation strategies.

I congratulate the authors for assembling a timely, coherent, and impactful contribution to the field of carbon dioxide capture during the tenure of their SPARC project. I am confident that it will stimulate further research partnerships and contribute to the development of sustainable technologies with global relevance.

Rabibrata Mukherjee, PhD
Professor, Department of Chemical Engineering
IIT Kharagpur, India &
National Coordinator, SPARC

***

It is with great professional pleasure that I write this note for the book *Carbon Dioxide Capture: Innovations in Advanced Material Interfaces*. The topic addressed in this book lies at the very heart of one of the most pressing scientific and societal challenges of our time: mitigating anthropogenic carbon dioxide emissions through sound, scalable, and scientifically rigorous solutions.

Over the past two decades, carbon capture research has evolved from a largely process-driven discipline into a deeply materials-centric science. What distinguishes this book is its clear recognition that the decisive breakthroughs in carbon dioxide capture will arise not merely from discovering new materials, but from understanding and engineering the interfaces within those materials. By placing interfaces at the centre of discussion, the authors adopt a perspective that is both intellectually mature and technically essential. Interfaces govern adsorption energetics, mass transport, selectivity, stability, and regeneration efficiency parameters that ultimately determine whether a material remains a laboratory curiosity or becomes an industrially viable solution. The authors of this book demonstrate commendable command over both fundamentals and front-line research. The chapters reflect a careful balance between theory and experiment, between molecular-level understanding and system-level relevance. Advanced porous solids, hybrid composites, membranes, and electrochemically active materials are not treated as isolated categories, but as interconnected platforms unified by interfacial phenomena. This integrative treatment is particularly valuable for young researchers, who often encounter these materials as fragmented topics across different journals and disciplines.

I am especially encouraged by the emphasis placed on durability, real-world operating conditions, and scalability. Too often, exceptional capture capacities reported under idealised laboratory conditions fail to translate into practical technologies due to interfacial degradation, moisture sensitivity, or energy-intensive regeneration. By explicitly addressing these issues, the authors signal a responsible and forward-looking research philosophy, one that aligns scientific innovation with societal needs. Equally noteworthy is the book's engagement with modern characterisation and modelling tools. The use of in-situ and operando techniques, combined with computational insights, reflects the current best practices in materials science. This approach not only deepens mechanistic understanding but also enables rational design strategies, reducing reliance on trial-and-error methods. Such rigour is essential if carbon capture technologies are to progress at the pace demanded by global climate targets.

From an academic standpoint, this volume will serve as a valuable reference for graduate-level courses and research groups working at the intersection of materials chemistry, physics, and chemical engineering. From a broader perspective, it offers clarity to readers seeking to understand how fundamental interfacial science can translate into tangible climate solutions. I congratulate the authors for their scholarly depth, clarity of vision, and commitment to addressing a problem of global significance. I am confident that this book will not only inform but also inspire future research in carbon dioxide capture and advanced material interfaces. It is a timely and meaningful contribution to the scientific literature.

Prof. (Dr). Gopal Mugeraya

Vice President (Technical Education) at NITTE (DU),

Former Director, NIT Agartala, NIT Mizoram and NIT Goa.

ACKNOWLEDGEMENT

This work represents the outcome of a sustained scientific journey, shaped by personal values, institutional support, and continuous engagement with the global research community. I, Santosh K. Tiwari, would like to begin by expressing my deepest gratitude to my parents, whose lifelong sacrifices, moral grounding, and belief in education laid the foundation for my academic pursuits. Their encouragement instilled in me a sense of responsibility to use science for societal benefit. I am equally grateful to my wife for her patience, resilience, and unwavering emotional support throughout the demanding phases of research, writing, and academic service. Her understanding and encouragement made it possible to pursue this work with focus and balance.

The scientific concepts explored in this book evolved over several years through persistent inquiry into carbon dioxide capture and advanced material interfaces. These ideas were further consolidated and refined during the organisation of the *New Materials in Carbon Capture and Environmental Remediation (NMCCER 2024)* conference, convened by the author at NMAMIT, Nitte. I gratefully acknowledge the SPARC Program under the Indo-UK Scheme (Project No. P3808) for its generous financial support, which enabled international collaboration, high-quality scientific deliberations, and substantial support toward publication-related expenses. This funding was instrumental in facilitating both the development and dissemination of the research outcomes. I sincerely acknowledge the visionary leadership and institutional support provided by Nitte (Deemed to be University). I express my profound gratitude to the Vice-Chancellor for creating a research-driven academic ecosystem that encourages interdisciplinary collaboration and international engagement. I am equally thankful to Prof. Praveenkumar Shetty, Research Director, for his consistent encouragement, strategic guidance, and efforts in strengthening research culture and global partnerships at the university.

I am deeply thankful to senior professors and distinguished researchers whose scholarly interactions, over many years, directly or indirectly shaped the intellectual framework of this book. In particular, I acknowledge the late Prof. Huzcko (University of Warsaw, Poland) for his exceptional scientific insight and lasting academic influence; Prof. Y. Q. Zhu (Exeter University, UK) for valuable discussions on advanced materials and nanostructures; Prof. Lola and Prof. Fang (University of Nottingham, UK), Principal & Co-Investigators, for their scientific leadership and continued support within the Indo-UK collaborative framework; Prof. Kuruvilla Joseph (IIST, India) for his broad perspective on sustainability and materials development; Prof. Ganesh Chandra Nayak (IIT Dhanbad, India) for interdisciplinary insights bridging materials, energy, and engineering systems; Prof. Sung Ha (Hanyang University, South Korea) for discussions on functional materials and interfacial science; and Prof. Luiz A. Ribeiro Junior (Institute of Physics, University of Brasília) for thoughtful exchanges on materials synthesis, processing, and characterisation. I also extend my sincere appreciation to the wider scientific community, whose published work, experimental data, theoretical frameworks, and critical analyses used directly or indirectly have contributed to the collective knowledge base upon which this book is built. SKT gratefully acknowledges Dr. Irfan, IIT Bombay, co-author of this book, for his substantial intellectual contributions, sustained scientific discussions, rigorous data collection, and dedicated efforts in managing copyright and publication-related matters. His continuous engagement was instrumental in ensuring the technical rigour and contemporary relevance of this work. Without his involvement, this book would not have been realised in its present form.

Finally, I gratefully acknowledge my PhD students, Mr. Madhusudhana M. Devadiga, Ms. Anushree S. Bhat, Abhishek R. Pandith and Subham Sarangi (NMAMIT, Nitte, India), for their dedicated assistance during the final stages of preparation, particularly in formatting and organisation. Their commitment and professionalism were invaluable. This book stands as a collective expression of personal support, institutional vision, and the cumulative efforts of the national and international scientific community striving towards sustainable carbon management and environmental remediation.

Finally, SKT offers all credit to Bhagwan Shri Sita Ram Ji Maharaj for His kindness, grace, and the opportunities bestowed upon him. SKT owns no credentials without His divine blessings.

AUTHOR'S REMARK

The escalating concentration of carbon dioxide (CO_2) in Earth's atmosphere is one of the defining scientific and technological challenges of the twenty-first century. Anthropogenic CO_2 emissions, primarily from fossil fuel combustion, cement production, and industrial processes, are the dominant drivers of global climate change, ocean acidification, and associated environmental disruptions. While renewable energy deployment and efficiency improvements are indispensable, they alone are insufficient to achieve deep decarbonisation within the required timeframe. Carbon dioxide capture (CDC), therefore, emerges not as an optional adjunct but as a critical pillar of a comprehensive climate mitigation strategy.

This book, *Carbon Dioxide Capture: Innovations in Advanced Material Interfaces*, is motivated by the recognition that the performance, scalability, and economic viability of CDC technologies are fundamentally governed by materials—specifically, by the interfaces within and between those materials. From gas-solid adsorption surfaces and liquid–gas absorption boundaries to electrochemical and membrane interfaces, it is at these nanoscale and mesoscale junctions that CO_2 molecules are selectively recognised, transported, transformed, or immobilised. Understanding and engineering these interfaces with precision is essential in overcoming the long-standing trade-offs between capacity, selectivity, kinetics, stability, and energy efficiency. Traditional CDC technologies, such as aqueous amine scrubbing, have demonstrated industrial feasibility but suffer from intrinsic limitations. High regeneration energy penalties, solvent degradation, corrosion, and environmental concerns motivate the search for next-generation alternatives. In response, the scientific community has developed a rich portfolio of advanced materials, including porous carbons, zeolites, metal-organic frameworks (MOFs), covalent organic frameworks (COFs), polymeric membranes, ionic liquids, hybrid nanocomposites, and electrochemically active solids. While these materials differ widely in composition and structure, their effectiveness converges on a common theme: the nature and behaviour of the material interface interacting with CO_2.

The central premise of this volume is that interfaces are not passive boundaries but active, tuneable regions where chemistry, physics, and mechanics intersect. Surface functional groups determine adsorption enthalpies and selectivities; pore apertures and connectivity regulate diffusion pathways; interfacial electric fields influence electro-swing and redox-mediated capture; and heterojunctions between dissimilar phases can generate synergistic effects unattainable in single-component systems. By focusing explicitly on advanced material interfaces, this book seeks to provide a unifying framework that links molecular-level phenomena to macroscopic capture performance. A distinguishing feature of contemporary CDC research is the increasing integration of interface engineering with predictive modelling and advanced characterisation. In-situ and operando spectroscopies, high-resolution microscopies, neutron and X-ray scattering, and surface-sensitive techniques now enable direct observation of CO_2–material interactions under realistic conditions. Complementarily, density-functional theory, molecular dynamics, and multiscale simulations provide mechanistic insights that guide rational design. This convergence of experiment and theory has transformed CDC from an empirical materials screening exercise into a knowledge-driven discipline grounded in interfacial science.

This book is structured to reflect both the diversity and the coherence of the field. Early chapters establish the fundamental thermodynamics and kinetics of CO_2 capture at interfaces, emphasising adsorption mechanisms, transport phenomena, and structure-property relationships. Subsequent sections examine major classes of capture materials, with particular attention to how interfacial modifications such as heteroatom doping, defect engineering, surface grafting, and composite formation—enhance performance. Dedicated chapters address membrane interfaces, electrochemical capture systems, and hybrid processes that couple capture with conversion or storage. Beyond performance metrics, durability and scalability are treated as first-order design criteria. Many promising materials exhibit exceptional laboratory-scale performance but degrade under moisture, oxygen, impurities, or repeated cycling. Interfacial instability, including chemical oxidation, hydrolysis, and mechanical fatigue, often dictates operational lifetimes. Accordingly, this volume emphasises interface stabilisation strategies, such as protective coatings, hierarchical structuring, and self-healing chemistries, that extend material longevity without compromising efficiency. Equally important is the techno-economic and environmental context in which CDC materials must operate. Interface-centric innovations must ultimately translate into systems that are cost-effective, energy-efficient, and environmentally responsible across their life-cycle. Several chapters, therefore, discuss process integration, regeneration strategies, and sustainability considerations, including material abundance, synthesis routes, and recyclability. By embedding interfacial materials research within a systems-level perspective, the book aims to bridge the gap between laboratory discovery and industrial deployment.

This volume is intended for a broad yet technically engaged readership. Graduate students and early-career researchers will find foundational concepts and state-of-the-art examples that clarify the role of interfaces in CO_2 capture. Established scientists and engineers will benefit from the synthesis of recent advances across traditionally siloed subfields. Policymakers and industrial practitioners with a scientific background may also draw insight into how targeted materials innovation can unlock new pathways for emissions reduction. The editors and contributors share the conviction that meaningful progress in CDC will arise from interdisciplinary collaboration. Chemistry, materials science, physics, chemical engineering, and computational science must converge to address challenges that are simultaneously molecular and societal in scale. Interfaces—by their very nature—symbolise this convergence. They are zones of contact, exchange, and transformation, mirroring the collaborative interfaces required among disciplines, institutions, and nations to confront climate change.

In presenting *Carbon Dioxide Capture: Innovations in Advanced Material Interfaces*, we do not claim to offer definitive solutions. Rather, this book is a snapshot of a rapidly evolving field, capturing both its achievements and its open questions. We hope it will stimulate critical thinking, inspire innovative research, and contribute, in however modest a way, to the global effort to manage carbon responsibly. The urgency of the climate challenge demands not only incremental improvements but conceptual advances, and it is at the interface, both scientific and collaborative, that such advances are most likely to emerge.

ABBREVIATION

CCS	–	Carbon Capture and Storage
CCU	–	Carbon Capture and Utilisation
CCUS	–	Carbon Capture, Utilisation, and Storage
CDC	–	Carbon Dioxide Capture
DAC	–	Direct Air Capture
DACU	–	Direct Air Capture with Utilisation
EOR	–	Enhanced Oil Recovery
IGCC	–	Integrated Gasification Combined Cycle
ASU	–	Air Separation Unit
MOF	–	Metal-Organic Framework
COF	–	Covalent Organic Framework
BAMC	–	Biomass-Activated Carbon Materials
CNM	–	Carbon Nanomaterial
CNMs	–	Carbon Nanomaterials
MEA	–	Monoethanolamine
SRM	–	Solar Radiation Management
PtX	–	Power-to-X
FESEM	–	Field Emission Scanning Electron Microscopy
SEM	–	Scanning Electron Microscopy
TEM	–	Transmission Electron Microscopy
XRD	–	X-ray Diffraction
IPCC	–	Intergovernmental Panel on Climate Change
SPARC	–	Scheme for Promotion of Academic & Research Collaboration
NMCCER	–	New Materials in Carbon Capture and Environmental Remediation
PTRC	–	Petroleum Technology Research Centre
BD3	–	Boundary Dam Unit 3
CSA	–	Canadian Standards Association
LCA	–	Life Cycle Assessment
NOAA	–	National Oceanic and Atmospheric Administration
IGCC	–	Integrated Gasification Combined Cycle
CO_2-EOR	–	Carbon Dioxide Enhanced Oil Recovery

HISTORICAL PERSPECTIVE

Carbon capture and utilisation (CCU) has evolved through a long and technically diverse history that predates modern climate policy. Initially emerging from advances in physical chemistry and industrial gas processing in the nineteenth century, CCU was practised implicitly in fermentation, lime calcination, and early ammonia production, where CO_2 was treated as an unavoidable by-product rather than an environmental liability. The development of the Bosch–Meiser urea synthesis process in the 1920s marked the first large-scale, intentional utilisation of CO_2, establishing its role as a viable chemical feedstock. From the 1930s to the 1970s, CO_2 utilisation expanded across food, chemical, and energy sectors, driven primarily by process efficiency and economic value. The commercial deployment of CO_2-enhanced oil recovery in the 1970s demonstrated large-scale CO_2 handling and transport, laying essential infrastructure foundations for later capture technologies. However, climate considerations were absent from these developments. A paradigm shift occurred during the 1980s and 1990s with growing scientific recognition of anthropogenic climate change, catalysed by assessments from the Intergovernmental Panel on Climate Change. This period prioritised carbon capture and storage (CCS), reframing CO_2 as a regulated pollutant, while utilisation remained secondary. In the early 2000s, CCU emerged as a distinct scientific discipline, integrating catalysis, materials science, and life-cycle assessment to evaluate CO_2 conversion pathways on a system level. The 2010s saw rapid diversification through mineral carbonation, electrochemical and biological conversion, and integration with renewable energy systems. In the current decade, CCU has become embedded within net-zero and circular carbon economy strategies, recognised by institutions such as the Global CCS Institute as complementary to permanent storage. Historically, CCU has transitioned from opportunistic industrial practice to a strategically designed climate technology, whose future impact depends on energy integration, carbon permanence, and scalable deployment.

Scientific and Industrial Origins of CO_2 Capture (19th Century to 1920s): The origins of CCU are deeply rooted in the development of physical chemistry and early industrial gas processing. During the 19th century, fundamental principles governing gas–liquid equilibria, adsorption, and reaction thermodynamics were established. Laws describing gas solubility and chemical equilibrium provided the intellectual framework for separating CO_2 from mixed gas streams. At this stage, CO_2 was neither perceived as environmentally harmful nor strategically important for climate considerations. Instead, it was viewed as a chemically stable, non-toxic gas with limited but practical industrial applications. CO_2 capture occurred implicitly in processes such as fermentation, lime calcination, and early fuel gas purification. These operations generated relatively pure CO_2 streams as unavoidable by-products, which were often vented or reused locally. A defining milestone in early CCU history was the industrial synthesis of urea in the 1920s through the Bosch–Meiser process. This marked the first large-scale, intentional chemical utilisation of CO_2. The process demonstrated that CO_2 could be incorporated into stable, value-added products using thermodynamically favourable reaction pathways. Importantly, this utilisation route remains dominant today, accounting for the largest fraction of global CO_2 use. This early phase established a crucial scientific insight: **CO_2 is not chemically inert waste, but a potential carbon resource under suitable energetic and catalytic conditions.**

Growth of Industrial CO_2 Utilisation and Process Integration (1930s to 1970s): Between the 1930s and 1970s, the industrial use of CO_2 expanded steadily, driven by process optimisation and market demand rather than environmental concerns. CO_2 found widespread application in food preservation, beverage carbonation, fire suppression, refrigeration, and chemical synthesis. These applications relied primarily on captured CO_2 from industrial processes, reinforcing localised capture–use loops. Technologically, this period saw incremental improvements in absorption-based capture systems. Alkaline solutions and early amine solvents were adopted to remove CO_2 from gas streams to meet purity specifications. Capture efficiency and energy consumption were not optimised for emissions reduction but for operational reliability. A major shift occurred in the 1970s with the commercial deployment of CO_2-enhanced oil recovery (EOR). Large volumes of CO_2 were injected into depleted oil reservoirs to improve extraction efficiency. While EOR demonstrated the feasibility of transporting and injecting CO_2 at scale, its objective was energy production rather than carbon mitigation. This era represents a **scale-up phase**, where CO_2 handling transitioned from small-scale chemical reuse to industrial-scale fluid management, laying critical infrastructure foundations for later CCS and CCU systems.

Climate Change Recognition and the Birth of Carbon Capture Technologies (1980s to 1990s): The scientific recognition of anthropogenic climate change in the 1980s fundamentally altered the perception of CO_2. Atmospheric measurements and climate models identified fossil fuel-derived CO_2 emissions as a primary driver of global warming. Institutions such as the Intergovernmental Panel on Climate Change formalised the link between emissions and climate risk. During this period, carbon capture emerged primarily within the framework of CCS. The emphasis was on preventing CO_2 release rather than using it as a feedstock. Capture technologies were developed for power plants and large industrial sources, including post-combustion amine scrubbing, pre-combustion gas separation, and oxyfuel combustion. Although utilisation remained secondary, this phase was essential for CCU development because it delivered high-purity CO_2 streams and techno-economic understanding of capture systems. The scale mismatch between global emissions and utilisation markets led to scepticism regarding CCU's mitigation potential, reinforcing storage as the dominant strategy. Thus, the 1980s to 1990s represent a conceptual transition, transforming CO_2 from a useful industrial gas into a regulated environmental liability.

Formal Emergence of CCU as a Scientific and Engineering Discipline (2000s): In the early 2000s, CCU began to emerge as a distinct research field at the intersection of chemistry, materials science, and energy systems engineering. CO_2 was reframed as a C1 building block, stimulating research into catalytic activation and conversion pathways. Advances in heterogeneous catalysis enabled efficient CO_2 hydrogenation to methanol and formic acid. Polymer chemists demonstrated the incorporation of CO_2 into polycarbonates and polyurethanes, challenging the perception that CO_2-derived materials must be low value. Simultaneously, mineral carbonation gained attention as a pathway for producing thermodynamically stable carbonates. A critical scientific contribution of this decade was the adoption of life-cycle assessment (LCA) as a standard evaluation tool. CCU processes were no longer assessed solely on conversion efficiency but on net carbon reduction, energy intensity, and product lifetime. This period established CCU as a systems problem rather than a single reaction, integrating capture, conversion, energy supply, and end-use considerations.

Diversification Through Materials Innovation and Renewable Integration (2010s): The 2010s marked rapid diversification of CCU technologies driven by renewable energy deployment and advances in materials science. Electrochemical and photoelectrochemical CO_2 reduction emerged as promising routes for coupling CO_2 utilisation with renewable electricity. These approaches offered modularity and compatibility with intermittent energy sources. Mineralisation pathways became matured, particularly in the cement and construction sectors, where CO_2 curing and carbonation provided both performance benefits and long-term carbon storage. Biological utilisation expanded through microalgae cultivation and synthetic biology, enabling the conversion of CO_2 into biomass and speciality chemicals. Importantly, CCU was increasingly classified based on **carbon permanence**, distinguishing short-lived products (fuels) from long-lived materials (minerals, polymers). This classification reshaped scientific and policy discussions, clarifying CCU's role alongside CCS.

CCU in Net-Zero Strategies and the Circular Carbon Economy (2020s to Present): In the current decade, CCU is embedded within net-zero emission strategies and circular carbon economy frameworks. Concepts such as Power-to-X integrate captured CO_2 with green hydrogen to produce synthetic fuels and chemicals. Direct air capture with utilisation extends CCU beyond point sources to atmospheric CO_2 removal. Organisations such as the Global CCS Institute now explicitly recognise CCU as complementary to geological storage, particularly for hard-to-abate sectors such as cement, steel, and chemicals. Modern CCU is evaluated holistically, considering scalability, energy source, material durability, and societal impact. The field has evolved from isolated chemical processes into a strategic component of climate mitigation portfolios. The summary of the overall progress is presented in tabular form.

Period	Dominant Driver	Key Technologies	Scientific Insight	Role of CCU
19th century	Industrial chemistry	Gas absorption, fermentation	CO_2 separable and reusable	Incidental by-product use
1920s	Agriculture	Urea synthesis	CO_2 as chemical feedstock	Large-scale utilisation
1930s to 1970s	Industrial efficiency	Absorption, EOR	CO_2 as process fluid	Scale expansion
1980s to 1990s	Climate science	Amine capture, CCS	CO_2 as pollutant	Capture-focused
2000s	Sustainability	Catalysis, polymers	System-level carbon balance	CCU formalised
2010s	Energy transition	Electrochemistry, mineralisation	Carbon permanence concept	Diversified pathways
2020s to present	Net-zero goals	PtX, DACU	Circular carbon economy	Strategic mitigation tool

Chapter 1

CARBON DIOXIDE CAPTURE: CURRENT STATUS AND FUTURE PROSPECTS

Abstract: As the world works towards net-zero emissions, the demand for large-scale CCS is growing. However, CCS implementation is still far behind what is needed to reduce greenhouse gas emissions significantly. Developing nations can more easily adopt and promote CCS by learning from the successful practices of developed nations. The development of major CCS infrastructure in different countries has been a key driver for technological innovation, cost reduction, risk management, and talent development in CCS technologies. In addition to those supportive policies, transitioning from pilot-scale projects to large-scale commercial infrastructures is crucial for the successful expansion of CCS. Without this, overcoming the technical challenges between small-scale trials and the establishment of CCS hubs with capacities of 1 million to 10 million tonnes will be difficult. The ultimate goal of CCS projects is geological CO_2 storage, but challenges remain in improving the accuracy of technology for measuring, monitoring, and verifying CO_2 storage capacity, emission reductions, and safety. One priority is CO_2 storage in saline aquifers, which can integrate various carbon emission sources. Key obstacles to widespread CCS adoption include reducing the energy required for capturing low-concentration CO_2, preventing the depletion of chemical absorbents, and improving the stability and efficiency of post-combustion CO_2 capture systems. Additionally, enhanced oil recovery is crucial for optimising fossil fuel extraction instead of relying on oil imports from less environmentally friendly sources.

1.1 Introduction

The continuous escalation of the CO_2 quantity (above 80%) in the atmosphere has posed a widespread threat to human life and thus has become a worldwide concern [1]. The main reason for this exponential rise in the CO_2 emission is mainly due to the ready use of fossil fuels, deforestation and industrialisation. Among the mentioned concerns, fossil fuels account for about 86% of the greenhouse gas emissions [2]. Since greenhouse gases possess adverse environmental effects, their reduction has gained national and international attention. Different initiatives have been taken nationally and internationally through government and different high-profile industrial persons, which have led to the development of platforms like the Intergovernmental Panel on Climate Change (IPCC), the United Nations Framework Convention on Climate Change, along with the global climate change initiatives. The main aim of these platforms is to develop the strategies through which the excess CO_2 can be converted into useful energy in different forms. This approach will not only help in reducing the CO_2 but will also facilitate the rising demands of energy from fossil fuels [3]. The initiative of CCS clubs involves multiple technological domains, such as its utilisation in the power plants, which involves multiple processes that are high compression, storage, and transportation.

Besides, the future strategy for curbing the threat and searching for complementary alternatives will be to focus on increasing the energy efficiency, switching on the renewable sources of the fuels such as solar energy, wind energy and biomass. The proposal of CCS is tedious and needs to be sophisticated, which can be attained through a common platform. This common platform needs not only recognition from the government but also direct efforts from different organisations, policy makers, economists, researchers, investors, and engineers. The mentioned aspects clearly provide an overview about the fact that why CCS is considered to be a grand challenge in the current century [3]. To have an overview related to the cost approximation of the proposal, many researchers have generated different reports by taking into account the engineering feasibility and cost of CCS. These reports have been generated on modelling post-combustion in the commercially available sources like coal and gas power plants [3, 4]. The presence of these sources dates back 50 years now. The pictorial representation of the progress that has been made from time to time in the CO_2 capturing is well depicted in **Figure 1.1**. The early attempts for CO_2 capture in these sources were based on separating the flue gases and the generated CO_2 by the use of scrubbers and some cryogenic cooling systems [5]. As per the estimations of the IPCC, it has been predicted that the modern power plants associated with the features of CCS will bring about an 80–90% reduction in the CO_2 emission [6] with the assumption that the total percentage of CO_2 flues present in the coal-based power plants is approximately 12-15% of total at 40°C. The minimum thermodynamic energy required for capturing 90% of the CO_2 generated by coal-based power plants has been roughly estimated to be 3.5% [7]. This technology has proved to be very efficient in many ways as compared to the capturing of CO_2 using scrubbers. While looking into the cost-effectiveness, for the mentioned estimates of capturing the scrubber-based technology increases the energy requirement of the power plants up to 25–40% [6–8]. Besides that, the short lifespan of the scrubbers will tend to enhance the cost of generated electricity by $0.06kWh [7]. This picture clearly brings our attention to the point that the technology of scrubbers is not cost-effective and demands technological progress [6].

Figure 1.1: The timeline representing the evolution and progress of the carbon capture, utilisation and storage technologies from the conventional approaches to the modern, modular, and low-energy systems. Copyright Credit: (Ref [7], Energy & Environmental Science, RSC, 2026)

The aspects of the cost-effective and energy-efficient open up a broad domain of research for setting up the stage for effective CCS technology. Different aspects of CCS, such as compression, transportation through the existing pipeline and their renovations, along with the storage, are simultaneously growing and progressing towards commercialisation. In this regard, throughout the world, different projects are being run in parallel in different countries, such as Sleipner West in Norway, Weyburn in Canada, Ketzin and Schwarze Pumpe in Germany and many more [8,9]. Apart from this, some multimillion-dollar projects have also been proposed by different researchers in different countries for attaining the goal of CCS. While looking deeply into the whole CCS process, the main hurdle in its progress is the expensiveness of the primary step, which is the capturing process, which approximately costs around two third of the total cost. For the mentioned process, there exist mainly two variables: regenerative energy and the cost of the employed equipment that needs to be made cost-effective [7]. In this regard, research is being carried out on maximising the separation efficiency of the capturing materials. This is one of the challenging aspects, and achieving efficiency in this will tend to reduce the cost of the overall system [4]. Before the system comes to the phase of commercialisation, we need to make progress in different aspects, as depicted in **Figure 1.2** [10]. For assessing the utilisation of any material, the primary step is to assess the design and cost of the material processing. As the optimisation of the material depends on several parameters, such as the type of the plant, fuel being employed, then comes its scale and production, which need to be balanced as per the norms of CCS [4,11]. Thus, from the above discussion, it is clear that CCS is an important aspect and thus need have proper insights about its need, current state and challenges. In this regard, this chapter will focus on the key challenges of the CCS technology. Implications on the human life by CO_2 that led to the concept of CCS. Some big projects, their progress timeline and outcomes. Finally, the chapter will be provided with a proper conclusion.

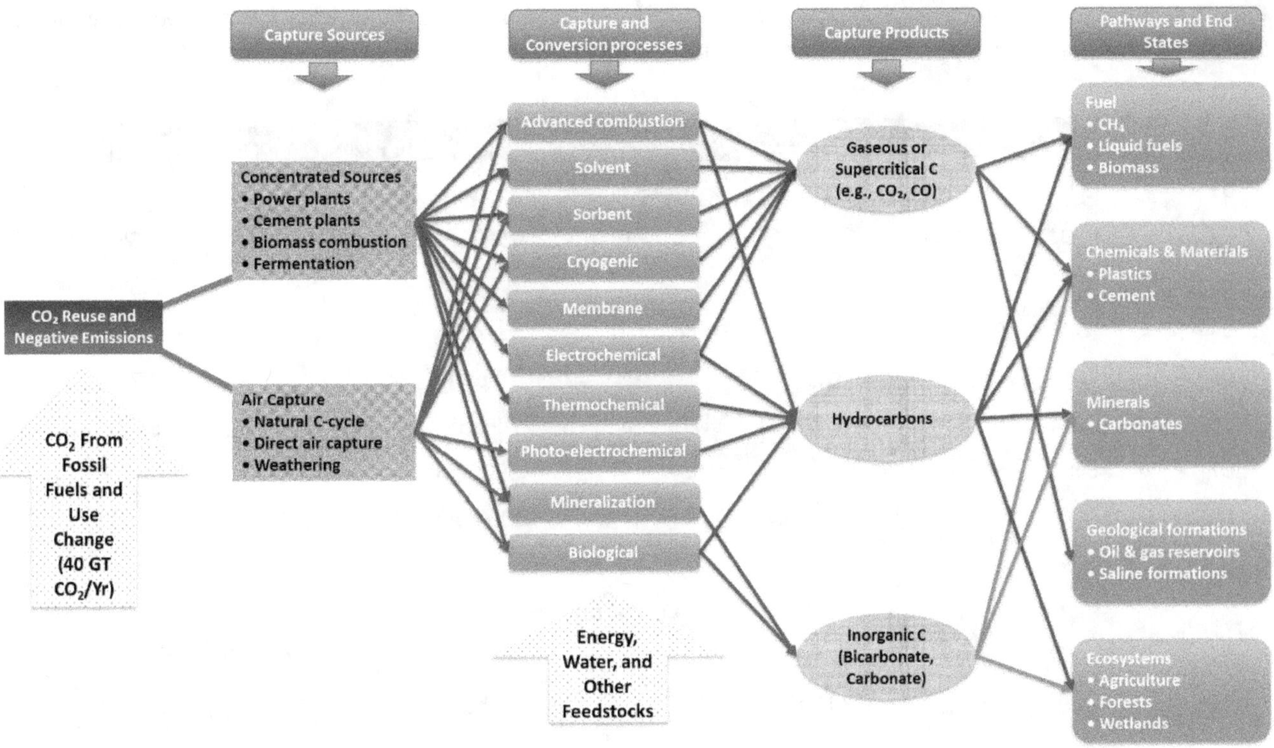

Figure 1.2: A thorough explanation of the variety of technology choices and methods for capturing, using, and storing carbon to reach net-zero or negative emissions.

1.2 Key Challenges for CO_2 Capture

Knowing the current scale of global CO_2 emissions, there are two key issues to consider when it comes to capture materials and technology [7]. First, if a chemical is used to capture CO_2 just once, its global supply will quickly run out. Second, if CO_2 is used as a reactant to produce a chemical, the market for that chemical will quickly become saturated. These challenges emphasise the importance of using regenerative materials for CO_2 capture. One of the main factors affecting both cost and efficiency in this process is the energy required for regenerating these materials. Among the different possibilities available, three processes are considered to be efficient, which possess the capability of reducing the CO_2 emissions. These include the separation of CO_2 from the power plant flue gases, sour natural gas wells, and fuel gas (or syngas). Each of these methods possesses its unique process for the separation of gas, which comes with its own specific material needs and limitations. Some of the useful benchmark criteria developed by the researchers for the post- and pre-combustion processes involved in the gas separation process are given in **Table 1.1** [12].

Table 1.1: Benchmark characteristics depicting the distinctive gas compositions (by weight) found in both pre- and post-combustion processes, along with the kinetic diameter of gas molecules.

S.No.	Composition	Pre-combustion (%)	Post-combustion	Kinetic Diameter (Å)
01	CO_2	35.5	15–16%	3.30
02	H_2O	0.2	5–7%	2.65
03	H_2	61.5		2.89
04	O_2		3–4%	3.45
05	CO	1.1	20 ppm	3.75
06	N_2	0.25	70–75%	3.64
07	SO_x		<800 ppm	
08	NO_x		500 ppm	
09	H_2S	1.1		
		Conditions		
01	Temperature	40°C	50–75°C	
02	Pressure	30 bar	1 bar	

The low pressure of flue gas represents a significant challenge for post-combustion CO_2 capture. As CO_2 makes up only about 15% of the gas, and its capture represents separating it from a large volume of exhaust gases, mainly nitrogen. Despite the fact that capturing through the post-combustion is appropriate for retrofitting with the existing technology, for different plants, we need to pay the energy penalty. The energy penalty for the coal-based plants which employ the conventional aqueous monoethanolamine (MEA) as capturing material has been approximated to be 25–40% [5, 7, 8]. Most of the gas reserves, especially based on methane (CH_4), consist of different harmful gases, of which the contributions of CO_2 and N_2 are approximately over 40%. The utilisation of such reserves is allowed only if they are accompanied by the capturing technology such that the excessive generated CO_2 will be captured and stored within the plant. However, for the implementation of effective CO_2 separation from the natural gas components, it is necessary that such plants are equipped with high-pressure systems.

Furthermore, in the pre-combustion process, the key separation is CO_2/H_2 and CO_2 separation from fuel gas, produced by gasification or water-gas shift reactors, can also occur at high temperatures (250–450°C) and under high pressure [11]. One of the biggest challenges in using materials for gas separation is that the gases we need to separate are very similar in their properties. This is reflected in their kinetic diameters (**Table 1.1**), which are defined as the minimum possible diameter of the molecule that needs to be measured. Although the gases have similar physical sizes, their electronic properties, such as quadrupolar moment and polarisation, differ more significantly. For example, consider three different molecules that are CO_2, N_2, and CH_4. Among these, the quadrupole moment of the CO_2 (13.4×10^{-40} cm^2) is much larger than that of the N_2 (4.7×10^{-40} cm^2), while the CH_4 molecule is non-polar. When considering adsorption, it has been observed that CH_4 possesses a much higher adsorption capacity than that of CO_2

and N_2. The higher adsorption capacity is associated with the polarisability values, which are 26.0×10^{-25} cm^3 for CH_4 compared to CO_2 (26.3×10^{-25} cm^3) and N_2 (17.6×10^{-25} cm^3) [13].

Therefore, to develop effective new gas capture methods, we need to control things at the molecular level, tapping into the subtle differences in how gas molecules react chemically. However, this presents another challenge: in most materials, the processes of adsorption and diffusion work hand in hand to determine how well a separation can be made. It has been observed that if we add a particular functional group that possesses an affinity for binding a particular type of molecule, thereby improving the adsorption capacity of that molecule. However, on the other hand, at the same time, the process of diffusion of molecules through the material gets reduced. Krishna *et al.* [14] have extensively studied this complex behaviour of adsorption and diffusion in different materials such as metal-organic frameworks, carbon nanotubes, and zeolites. Therefore, to address the mentioned challenge, we need to develop materials which possess the characteristics at the molecular level such that we can simultaneously fine-tune both the adsorption and diffusion processes. Another approach for reducing atmospheric CO_2 levels is to capture CO_2 directly from the air, especially from daily life sources like home furnaces, buses, cars, and other accessories [15]. However, the task is more difficult because CO_2 is present in very low concentrations in the air (around 0.04%), which creates a much tougher thermodynamic barrier for capture compared to post-combustion methods. Another challenge is the cost of moving large volumes of air through a material that absorbs CO_2. There's also ongoing debate about the actual costs of these methods, which also need to be taken care of for attaining the success. However, from the research standpoint, it's important to explore how direct air capture and CO_2 capture from flue gases can work together, especially considering the uncertainty about what the ideal levels of CO_2 in the atmosphere will be in the future [11].

1.3 Impact of CO_2 on Earth and Life

An irrevocable climatic catastrophe is imminent. There is no question that this is a worldwide emergency. The entire foundation of life on Earth is in danger. A crucial and unpredictable new stage of the climate catastrophe is upon us. Researchers working in different domains, particularly a consortium of over 15,000 scientists, have been warning about the looming threats of climate change caused by rising greenhouse gas emissions and ecosystem change for many years [16]. Due to advancements in technology, the alarming situation of global warming has been accurately anticipated by research scientists and even by many business entrepreneurs before its arrival [17]. Despite knowing the primary arrival and its bad consequences, we still keep going in the wrong direction without taking any proper measures to save the planet's ecosystem. Due to the progressive rise in the emission of ecosystem toxic gases from the continuous utilisation of fossil fuels, we experienced the three highest temperatures for three consecutive days in July 2024 [18]. It is expected that, if necessary, measures are not taken, we will experience global peak warming with an approximate rise in temperature of 2.7°C by 2100 [19]. All the mentioned are quite pictorially evident from the plotted data from NOAA, National Centre of Information, and the report presented by the Global CCS Institute for 2025 **(Figure 1.3)**.

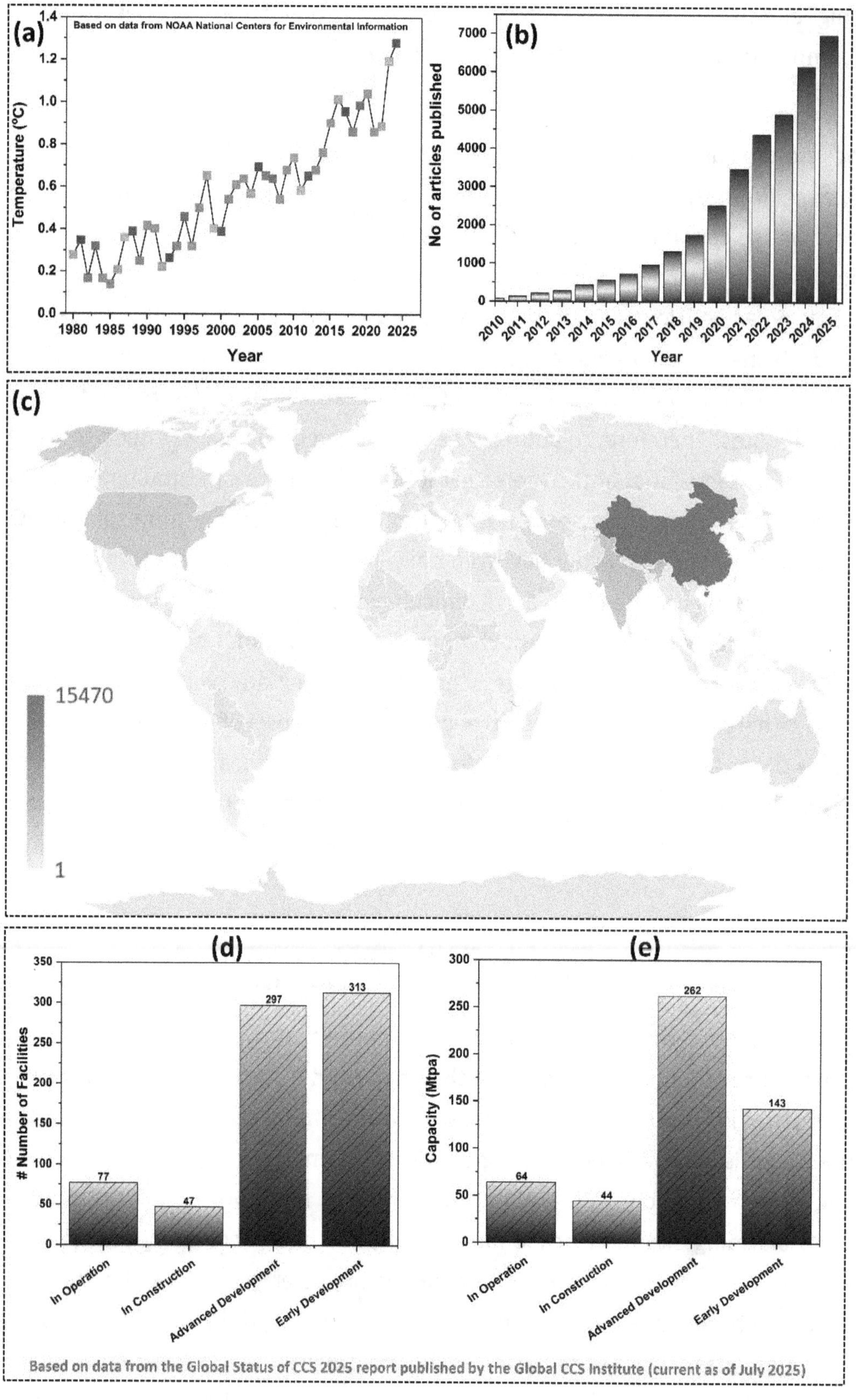

Figure 1.3: (a) Trend depicting the increase in the surface temperature as time progresses (as per NOAA data), (b) Data extracted from the Scopus (past 15 years) using the keywords "CO_2 Capture using the carbon materials" depicting the upsurge of research in this field, (c) Global publication record presenting the essence of research being carried out on CO_2 capture and sequestration, (d, e) growth and expansion in the facilities for CCS at different operational, construction as well as advanced development stages. Copyright Credit: (Ref [12], Materials Advances, RSC, 2026)

Unfortunately, we are unable to prevent significant effects, and at this point, our only option is to try to minimise the harm. As the effects of climate change intensify, we are seeing the sobering reality of the predictions, which are bringing about scenes of unheard—of global catastrophes and misery for both humans and other species. We are in the midst of a sudden climatic crisis, a grave circumstance that has never been seen in human history. We have now changed the planet's climate to one that our ancient genus members have never experienced [20]. The circumstance is so adverse that last year we experienced many uneven climatic changes which have not been observed in the past 2000 years. For the first time in history, the highest ever sea surface temperature has been witnessed in the northern hemisphere [21]. Besides these, many other climate changes have been observed [22, 23]. In addition, it is believed that the upcoming years will witness a significant increase in the harsh weather [24]. The primary cause of all these drastic climate changes is the generation of greenhouse gases, among which the most dangerous and the primary responsible is the CO_2 gas, generated through different human activities. While analysing the reason till 2022, it has been observed that around 90% of these emissions are caused by industrial operations and the burning of fossil fuels worldwide, while changes in land use cause 10%, mostly deforestation [25]. All these climate changes have affected the normal life on the planet through extreme climate and weather changes, melting of water reservoirs, variation in the solar radiation, etc. Some highlights about these areas are:

1.4 Effect of Climate Changes

Extreme weather and calamities brought on by climate change have significantly improved the human misery. Recent years have experienced an ever-increasing heat and precipitation rates, which have not been evidenced before in the climate history [26]. The rise in global temperature due to excessive emission of greenhouse gases has resulted in the generation of extreme heat waves. This has led to an enormous number of consequences associated with life on the planet, such as an increase in the mortality rate, hike in the healthcare systems, issues about mental health, as well as fatalities due to cardiorespiratory disease [27]. Considering only the United States, the increase in the fatality rate pertaining to the heat issues from 1999 to 2023 has increased significantly to 117% [28]. Besides that, the United States tied for the record with four floods last year, which caused damage of about 4 billion dollars. Apart from this, many other climate-related disasters have occurred, including strings of heat waves that killed over 1000 people in Asia, along with the tragic temperature rise to 50°C in different parts of India [23]. Owing to the nonlinearity in the Earth and its weather systems, the catastrophic changes will affect both plant and animal life [26, 29].

The climate changes primarily affects the coral reefs, which act as a source of livelihood for different people apart from being a habitat for many species [30]. One of the major risks is coral bleaching, where corals lose their symbiotic relationship with microalgae, often leading to their death [31]. Sometimes, a sudden spike in water temperature, much higher than the usual summer maximum, can kill corals within just a week or two, without giving them time to bleach. Other times, corals bleach but may recover their colour and survive, or they die more gradually over a few months [32]. As of 2024, the world is experiencing its fourth global-scale coral bleaching event, driven by unusually high ocean temperatures [33]. Climate change is also altering the hydrology and biogeochemistry of Arctic watersheds. In recent years, researchers have observed an emerging threat in Arctic streams that have turned orange due to higher levels of harmful metals and iron. This discolouration began within the last decade, as permafrost melting and intensification of global warming [34]. The orange streams have higher turbidity, sulphate, iron, and trace metals, and are more acidic compared to clear streams. This change is affecting drinking water quality and subsistence fishing in rural Alaska, and it's linked to declines in fish populations and a reduction in the diversity of macroinvertebrates [34].

1.5 Variation in Solar Radiations

Research into solar geoengineering, or solar radiation management (SRM), has picked up a race from the past many years. SRM aims at reflecting sunlight away from earth for reducing the impacts of climate change. The techniques employed by the SRM for attaining the aim are considered to be very risky. These methods can be applied in different areas, including the atmosphere, on land, and even in space [35]. Among the different techniques, some of the specific techniques being explored include brightening marine clouds and injecting aerosols into the stratosphere. According to the Intergovernmental Panel on Climate Change (IPCC), the most studied SRM method is stratospheric aerosol injection. However, there are numerous environmental concerns associated with these approaches [2, 36]. Apart from these concerns, SRM is also criticised in view of some ethical issues and other factors. Critics argue that relying on these methods could be risky, potentially diverting attention from efforts to reduce emissions and manage weather patterns more responsibly [37]. Research on the viability and risks of SRM is still ongoing. SRM is often seen as a potential short-term solution to decrease the warming and its associated impacts, especially in subpolar regions, which are warming rapidly and are of significant concern [38]. Future research should focus on understanding the possible environmental, societal, and political effects, while also evaluating the safety and effectiveness of solar geoengineering at both regional and global levels [39]. Additionally, there is a need for interdisciplinary studies to explore governance, legal, and ethical frameworks, as well as public acceptance and perception. It's also crucial to emphasise the urgent need to dramatically cut greenhouse gas emissions [24].

Thus, looking into the different aspects mentioned, it is evident now that the greenhouse effect, the main component of which is the CO_2 gas, severely affects life on the planet. It broadly affects every sector of life, starting from plants, aquatic life, humans, and other forms of life. Thus, in order to save life and future generations from drastic catastrophes, it is necessary to take special measures to convert CO_2 emissions into some useful form of energy.

1.6 Overview of Different Approaches Used for Carbon Dioxide Capture

The upgradation of the available power plants through the post-combustion capturing (**Figure 1.4 (a)**) techniques is currently the most commercially viable option, with two main alternatives [40,41]. However, it is believed that the new expensive developing technologies for capturing and processing, which are: pre-combustion and oxyfuel, as depicted in **Figure 1.4 (a)**, will eventually offer greater efficiency in CO_2 separation and capture, thus balancing the higher initial capital costs [6]. In natural gas facilities, pre-combustion techniques are used for CO_2 capture at high partial pressures [8, 40, 42]. The three-stage Integrated Gasification Combined Cycle (IGCC) facilities have been developed using pre-combustion methods to convert methane-rich fuel into hydrogen. In this process, firstly, natural gas is allowed to react at high temperatures and pressures to produce syngas, which contains carbon monoxide (CO), carbon dioxide (CO_2), and hydrogen (H_2). In the second step, CO is converted by employing the water-gas shift reactor, which separates CO_2 and H_2 before the initiation of burning. The reactions involved are given as:

$$H_2O + CH_4 \rightarrow 3H_2 + CO$$

$$H_2O + CO \rightarrow H_2 + CO_2$$

Although the initial fuel conversion steps in pre-combustion systems are more complex and costly compared to post-combustion systems, they offer significant advantages for CO_2 separation [42]. The shift reactor produces CO_2 at much higher concentrations, typically 15–60% by volume (dry basis) and at higher pressures, which makes the separation process more efficient. Another benefit is that these higher concentrations and pressures help in reducing the energy penalty for CO_2 capture, thereby lowering it to 10–16%, which is about half of what is required for post-combustion CO_2 capture. Additionally, pre-combustion technology produces a hydrogen-rich fuel, which can be used in various applications, such as a chemical feedstock, for power generation in fuel cells, or in the future progress of hydrogen. However, different initiatives have been taken for commercial IGCC facilities with Carbon CCS through the inauguration of different projects. The initial projects include China's 250 MW GreenGen project, along with the 275 MW FutureGen facility in the United States [8]. In another process called oxyfuel combustion depicted in **Figure 1.4 (a)**, nitrogen is completely removed from the combustion process through a preliminary air separation phase [40]. This occurs in a pressurised CO_2-rich recirculating stream, unlike post- and pre-combustion methods. Fuels such as coal, natural gas, light-to-medium hydrocarbons, and syngas (which can be derived from coal) are all suitable for use. The combustion process produces a concentrated CO_2 stream, with sequestration concentrations of the order of 80–98%, which is then sent back into the system through a proper channel. Besides that, the removal of water takes place through the condensation processes [11].

This technology is the building block for the concept of plants with a "zero-emission cycle." These plants are considered to be more appealing for the new setups compared to post- and pre-combustion capturing technologies for CO_2. This essence is clearly evident from the comparison made on the different aspects like driving forces, specific energy demands, efficiency, and CNMs. integration potential, presented in tabular form as **Table 1.2**.

Table 1.2: Comparative assessment of different CO_2 capturing technologies, in terms of thermodynamic driving forces, energy requirements, efficiency, and integration potentials.

Metric	Pre-Combustion Capture	Oxyfuel Combustion	Post-Combustion Capture
Thermodynamic Driving Force	**Pressure-driven separation:** High CO_2 partial pressures result in low-energy physical separation processes.	**Phase-change–driven separation:** Near-pure CO_2 streams allow condensation-based capture.	**Chemically driven separation:** Diluted streams of CO_2 demand enhanced chemical formation and cleavage.
Specific Energy Demand	**~2.5 MJ kg^{-1} CO_2 (lowest):** Syngas generation, along with CO_2 compression, is the main phenomenon associated with energy.	**~3.0-4.0 MJ kg^{-1} CO_2:** Energy consumption is primarily driven by the electricity demand of the air separation unit (ASU).	**~3.5-4.5 MJ kg^{-1} CO_2: (highest):** Substantial steam input required for solvent regeneration (reboiler duty).
Efficiency	**High:** CO_2 separation occurs upstream of expansion, minimising irreversibility and exergy destruction.	**Moderate:** Significant combustion-related energy losses, partially mitigated through latent heat recovery.	**Low:** High entropy generation arising from CO_2 separation from nitrogen-diluted flue gas.
CNMs. Integration Potential	**Transformative:** Membrane reactors enable simultaneous reaction and separation, shifting equilibrium, and eliminating intermediate cooling steps.	**High:** Membrane-based oxygen separation can replace cryogenic ASUs, substantially reducing parasitic energy losses.	**Incremental to moderate:** Advanced solid sorbents reduce sensible heat penalties, but the fundamental entropy constraint remains.

But for the implementation, the process necessitates redesigning the existing technology of turbines, which demands extra cost and thus makes it unfeasible to upgrade the currently working plants. In this regard, the first successful attempt for the establishment of the CO_2 capture plant employing the Oxyfuel technology was made in 2008 in Germany by the European energy firm called Vattenfall [43]. The air separator unit's high-energy requirements are one of the process's drawbacks, although the removal of the requirement for CO_2 capture should offset this expense. However, the unit was found to be accompanied by one major drawback: high-energy requirements in the unit employed for separating the air. The rising energy can be reduced by reducing the excessive need for the CO_2 capture. These approaches are further subdivided into different categories depending on the capture pathway and employed technique **(Figure 1.4 (b))**.

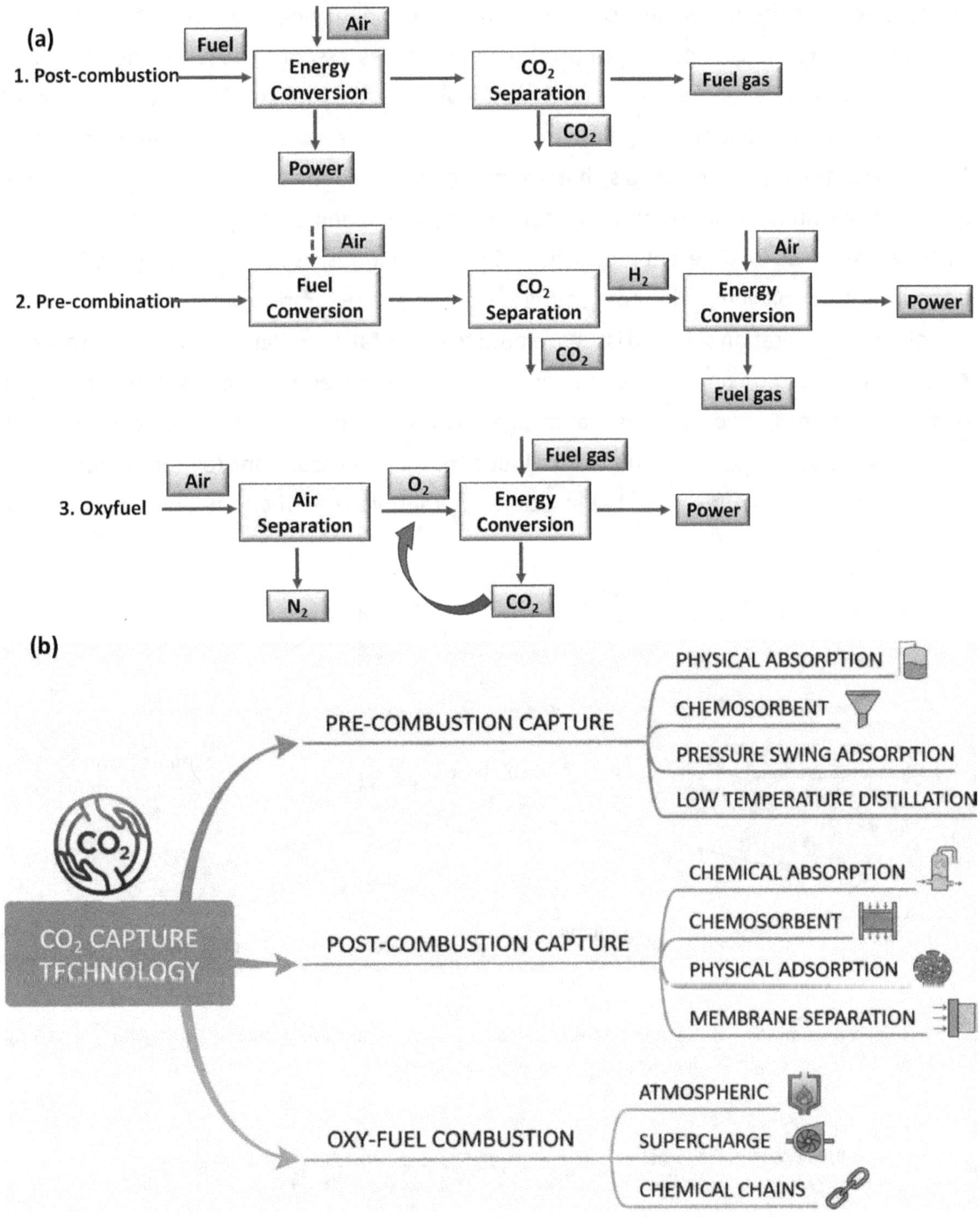

Figure 1.4: (a) Schematic representation of the different steps involved in the post, pre and oxyfuel combustion processes. (b) Different pathways and capturing technologies involved in the post, pre and oxyfuel combustion processes. Copyright Credit: (Ref [11], Angewandte Chemie International Edition, Wily, 2010 & Ref [12], Materials Advances, RSC, 2026)

Also, the efficiency of the boiler unit can be enhanced by managing the amount of nitrogen gas in the furnace. Lowering the cost of oxygen production becomes essential because the main separation process here is air separation, which involves extracting O_2 from mostly N_2. In this context, oxygen-selective membranes present an intriguing new technique as a successor to cryogenic air isolation [44]. While looking into the type of materials that are being employed for the mentioned techniques, there exists an enormous number of promising materials that have been employed for CO_2 capturing so far. These materials are used for different purposes in these techniques, as depicted in **Figure 1.5**. Traditional solvent absorbents employed in the post-combustion flue gas separations are known for their energy-intensive solvent regeneration step and issues related to chemical degradation. As a result, there has been a lot of research focused on finding better alternatives. Novel materials being explored include physical absorbents, membranes, cryogenic distillation, gas hydrate formation, chemical-looping combustion using metal oxides, and solid adsorption techniques that use pressure and/or temperature swings [45]. These innovative methods offer potential solutions to improve the efficiency and sustainability of gas separation processes.

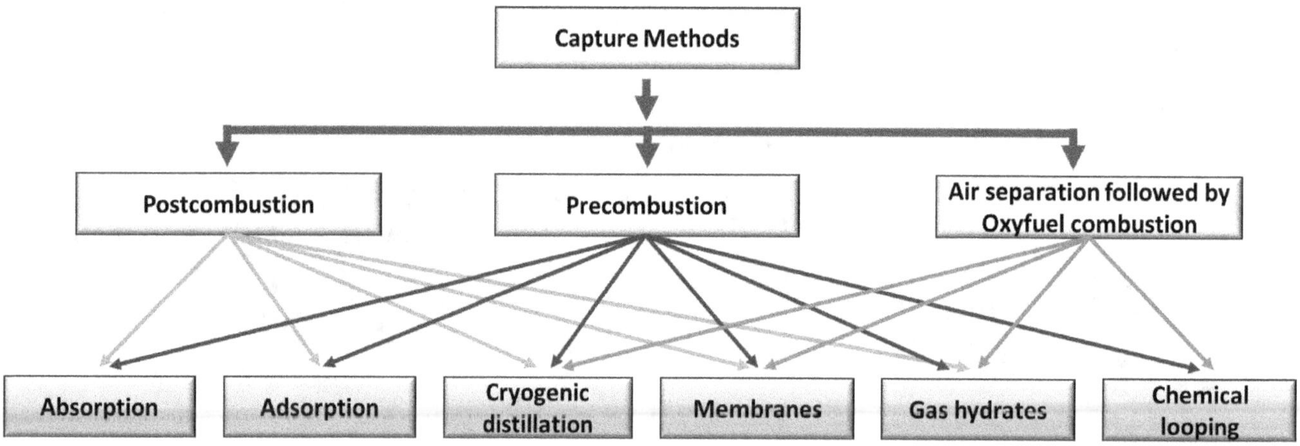

Figure 1.5: Pictorial representation of the materials employed for CO_2 capturing about the pre-combustion, post-combustion, and Oxyfuel operations. Copyright Credit: (Ref [11], Angewandte Chemie International Edition, Wily, 2010)

1.7 Key Global Companies Working on Carbon Dioxide Capture, Commercial Aspect

The concept of enhanced oil recovery (EOR) has a long history, dating back to the 1960s in the United States and Canada. However, it wasn't until January 27, 1972, that the first large-scale EOR project began, known as the Scurry Area Canyon Reef Operating Committee (SACROC). This project was spearheaded by the American multinational corporation, Chevron, and established in Scurry, Texas [46]. The main source of CO_2 for the project came from natural CO_2 found in the Colorado fields. Additionally, pipeline systems were put in place to increase the fuel content from the oilfields. By 2009, the SACROC project had processed approximately 175 million tonnes of natural gas since its inception. It's important to note that at the time, there was no focus on reducing the environmental impact of CO_2.

The project was aimed at generating an alternative version of fuel for everyday use [47]. The concept of CCS, designed to mitigate the harmful environmental effects of anthropogenic CO_2, was first introduced by Marchetti [48]. Today, CCS, also referred to as Carbon Capture, Utilisation, and Storage (CCUS), is recognised worldwide as a vital tool for combating climate change and protecting the ecosystem, as shown in **Figure 1.6(a)** [49, 50]. The first international CCS project began in 1996, known as the Sleipner CCS project, followed by the Weyburn Project 4 years later. The first research project aimed at commercial CO_2 production was carried out by Equinor, under Norway's carbon tax policy. This project aimed to reduce CO_2 levels in the atmosphere by separating CO_2 from natural gas in the first step and storing it in saline aquifers in the second. Since 1996, the project has stored nearly 20 million tonnes of CO_2. Over nearly two decades, the project has been closely monitored, producing significant scientific insights [51]. The Weyburn project, which began in 2000, is one of the most successful CO_2 storage projects. For almost 14 years, it has successfully stored CO_2 at a rate of 1.8 million tonnes per year, accumulating around 35 million tonnes of CO_2. This project is located at the world's largest geoscience test site, in the Weyburn fields of Canada. The project was funded by a combination of Canadian Natural Resources, the US Department of Energy, various government agencies, and private enterprises [52, 53]. After achieving success in its research objectives, the project quickly transitioned into a commercial endeavour. The achievement of the Weyburn Project highlighted the crucial role that CCS technology plays in decreasing emissions and thereby reducing the effect of greenhouse gases, the roadway for which is depicted in **Figure 1.6(b)**. It resulted in the establishment of Dakota Gasification Company, which became the world's first initiative to capture and transport high-concentration CO_2 from coal consumption on a large scale, over long distances, for EOR and storage. With this achievement, it is believed that CCS technology can be effectively employed for decreasing the emission of CO_2 from different fossil fuel sources, such as coal, one of the largest contributors to CO_2 emissions. The parametric features, such as performing the operation on a commercial scale, short time span, and cost-effective have made this a key solution for efficient usage of coal as fuel [47]. What sets this project apart is its independence; rather than relying on government funding, it has remained successfully depended on CO_2-EOR for over 22 years. This long-term success has made it one of the standout achievements in CCS technology. After overcoming challenges related to the effective and efficient capture of CO_2 from major emission sources like industries, the Weyburn project turned its focus to capturing CO_2 from lower-concentration sources, such as coal-based power plants. One of its key successes in this regard was the development of a post-combustion CO_2 capture unit (with a capacity of 1 Mta^{-1}). This unit was installed at the Saka Power Boundary Dam Power Station, Unit 3 (BD3), to facilitate the easy transport of captured CO_2 to the oil field for enhanced oil recovery and storage [47].

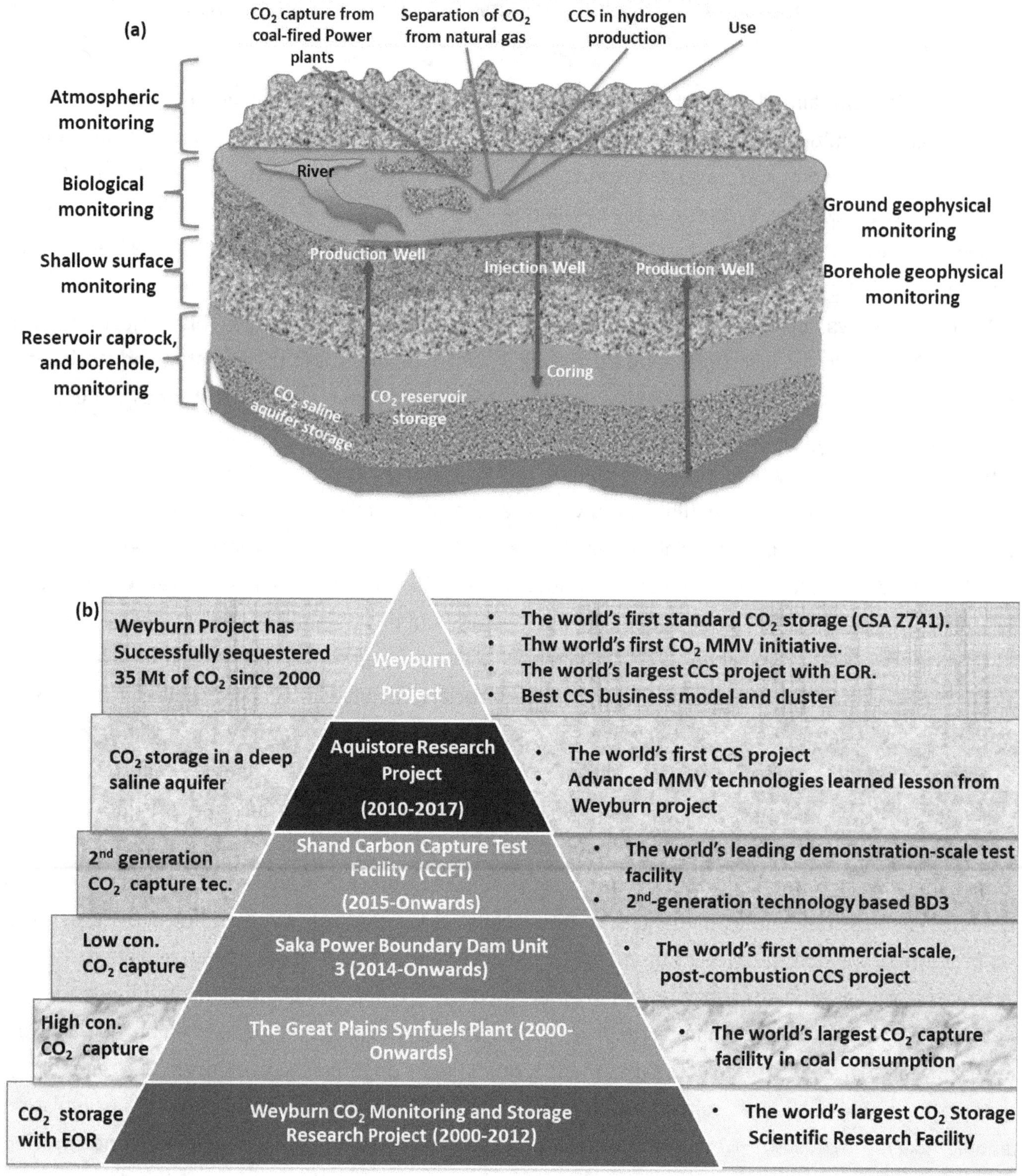

Figure 1.6: Pictorial representation for the (a) Industrial chain for CCS along with its different contributing factors (b) Weyburn Project's achievement made possible by strong scientific research and research facilities: BD3: SaskPower Boundary Dam Power Station Unit 3; CSA: Canadian Standards Association [47]. Copyright Credit: (Ref [47], Engineering, Elsevier, 2022)

The project is designed to handle fluctuations in demand, with any excess CO_2 being stored in saline aquifers when necessary. To achieve this, Saka Power, in collaboration with the Petroleum Technology Research Centre (PTRC), developed Aquistore, a facility where excess CO_2 is safely stored in saline aquifers. One of the groundbreaking successes of the Weyburn project is its ability to utilise CO_2 on a large scale without relying on government intervention or aid, making it one of the best commercial models for CCS development. This addition of "Utilisation" (the "U" in CCUS) has been widely praised around the world. Although a lot of progress has been made in CCS, some of the targets set by different nations to reduce the effects of CO_2 emissions on a large scale have not yet been achieved; further progress is needed. In this regard, different projects are being put forward that work on various aspects. In this section, a brief overview related to the categorisation of different CCS projects will be provided. The projects related to CCS have been categorised on different bases; however, the most well-known and globally acclaimed is the adapted Global CCS Institute (GCCSI) [54]. The threshold criteria set by the GCCSI for the testing and commercialisation of CCS projects rely on the capture source: with industrial emission sources, GCCSI accepts CCS projects that can store at least 40 $kt.a^{-1}$ of CO_2, while for emission sources like coal power plants, only projects with CO_2 storage capabilities of around 800 $kt.a^{-1}$ are accepted. According to the 2020 reports of GCCSI, they have accepted a total of 65 commercial projects globally, of which 26 are operational, while the rest are in different phases of progress and development. Another classification is based on the source responsible for CO_2 emissions, such as natural gas separation, coal chemical plants, oil refining, coal power plants, steelwork plants, and biomass units. For these sources, CCS projects like Sleipner, Great Plains Coal Gasification Plant, Shell Quest, Tomakomai, SakaPower Boundary Dam, Petre Nova, Al Reyadah, and Decatur have been accepted. However, projects that employ CO_2 for inundation and storage from natural CO_2 sources are not considered in the projects working on CO_2 reduction and are therefore not regarded as CCS projects [47]. Additionally, some of the leading companies working on different CCS projects globally are presented in **Table 1.3**.

Table 1.3: Different companies working for the CO_2 capturing and utilisation, with their average capacities, capturing methods, and the innovative area.

Company	Country	Focus Area	Key Technology	Keynotes
Company	Switzerland	Direct Air Capture (DAC)	Orca, Mammoth (Iceland)	Up to 36,000 t y^{-1} (expanding)
Climeworks	Canada	DAC + Air-to-fuel	Permian Basin DAC plant with Occidental	Target: 1 Mt y^{-1} (under development)
Carbon Engineering	USA	DAC + geological storage	Oxy 1PointFive, DAC hubs	Strategic DAC investments
Occidental (Oxy)	Norway	Modular industrial CO_2 capture	Just Catch™ technology for cement & waste plants	Deployable units <100,000 t y^{-1}

Company	Country	Focus Area	Key Technology	Keynotes
Aker Carbon Capture	UK	Modular CO_2 capture	CycloneCC modular technology	0.33 Mt y^{-1} (2022), cost-effective
Carbon Clean	Norway	Complete CCS chain, offshore storage	Sleipner, Northern Lights	Sleipner: >1 Mt y^{-1} since 1996
Equinor	France	Industrial CCS & hydrogen	Leverages chemical-looping combustion	Various projects in the EU
TotalEnergies	Germany	Cement CCS	Brevik CCS plant in Norway	400,000 t y^{-1} from 2024
Heidelberg Materials	Germany	Advanced solvents	Ionic liquid-based absorption systems	Ongoing industrial pilots
BASF	USA	Bioconversion of CO_2/CO	Microbial gas fermentation (ethanol, fuels)	Multiple plants, ~150,000 t y^{-1}
LanzaTech	Iceland	Mineralisation in basalt	CO_2, + water → carbonates in subsurface basalt	Permanent storage with DAC integration
Carbfix	USA	Industrial CCS	Gorgon project (Australia)	3.4–4 Mt y^{-1} (underperforming initial target)

There are several methods used for CO_2 storage in CCS projects, such as storage based on saline aquifers, EOR based storages, decommissioned oil and gas fields, as well as storage based on basalt formations [55]. Two notable examples of basalt CO_2 storage are the Carbfix and Tomakomai projects in Iceland and Japan, respectively [56, 57]. Although the Tomakomai project attained success in the insertion of CO_2 in sandstones, the efforts made to store was successful in completely injecting CO_2 into sandstone; efforts to store CO_2 in basalt have not proved to be as effective. This is largely because the Tomakomai reservoir lacks the high-temperature conditions and injectivity that are present in the basalt of the Carbfix project. As a result, mineralised CO_2 storage in basalt hasn't been achieved yet, suggesting that basalt may not be the most promising option for CO_2 storage at this time. Classifying CCS projects based on their scientific research and commercial demonstration or operation stages can give us valuable insights into their development journey, the challenges faced, and the technical progress achieved. As mentioned earlier, despite the extensive use of CO_2 flooding in the United States since 1972, CO_2–EOR technology focuses on increasing oil recovery, where a portion of the injected CO_2 remains in the reservoir, and the rest exits with oil from the producing wells. However, this approach is not primarily focused on the geological storage of CO_2 or on reducing CO_2 emissions. One crucial question that arises when CO_2 flooding technology shifts toward CO_2 geological storage, whether in oil reservoirs or saline aquifers, is: Is the CO_2 stored safely? How can we be certain about where the stored CO_2 actually should be stored? How do we verify the capacity of different storage sites to hold CO_2 safely? What tools and technologies will be employed to look into the long-term safety of CO_2 storage? These important questions form the backbone of making CCS commercially viable, and they need to be thoroughly addressed by research and infrastructure efforts in this field. One example is the Weyburn Project, a scientific initiative focused on carefully monitoring and storing CO_2, with extensive testing to ensure safety and effectiveness [52, 58].

The research from this project laid the groundwork for the Canadian Standards Association's (CSA) Z741, which is a standard on the geological storage of CO_2. This, in turn, became the foundation for the International Organization for Standardization's (ISO) or Technical Committee 265 standard for CO_2 capturing, shipping, and storage. Furthermore, research on storing large amounts of CO_2 underground is happening offshore through the Sleipner project. This project uses time-lapse gravity and 4D seismic monitoring to track how CO_2 moves and builds up beneath the surface during storing. Other important international CCS sites include Aquistore, CO_2 CRC Otway project, Tomakomai, Ketzin, and the Decatur project, which is notable for being the first CCS project involving biomass (**Table 1.4**) [59, 60]. The advancement of science and technology has resulted in the development of facilities covering all stages of CO_2 storage: from studying the site before injection, to monitoring during the injection process, and continuing research afterwards, even after the site is closed. They use advanced construction methods along with sophisticated monitoring tools and equipment. Early examples of such research sites include the K12-B Offshore CO_2 Injection Project in the Netherlands and a geological CO_2 storage test site in Nagaoka, Japan [47].

In general, all the CCS technological projects mainly focus on assessing, monitoring, and verifying how well CO_2 can be stored underground. They also test different monitoring technologies for storage sites that handle anywhere from 10,000 to 1 million tonnes of CO_2 per year. However, a few projects also focus on storing CO_2 in saline aquifers because these underground saltwater reservoirs can hold much more CO_2 than oil fields. The exception is the Weyburn Project, which looks at storing CO_2 alongside EOR. To keep costs down, especially for long-distance CO_2 pipelines, many projects store CO_2 close to where it's produced. The CO_2-EOR shouldn't be overlooked because it generates revenue that helps offset some of the high costs of capturing CO_2. Technically, when storing CO_2 in saline aquifers, the process involves directly injecting CO_2 without removing any fluids to decrease the in-situ pressure. This is different from CO_2 flooding in oil and gas reservoirs, where CO_2 is injected to maintain pressure and push out oil. When CO_2 is injected into saline aquifers, it can increase the in-situ pressure beyond the natural levels. This higher pressure might cause cracks in the caprock, fractures in the reservoir, and other faults, thereby creating potential risks of leakage. However, measures are being taken for managing these risks effectively, by ongoing international CO_2 storage projects such as Sleipner, Shell Quest, Tomakomai, Aquistore, and CO_2–CRC Otway [47].

Table 1.4: Different CCS technological projects, their structural information, and characteristics.

S.N	Project name	Source and scale	Transportation	Storage System	Approx. Amount of CO_2 Stored	Characteristics features
01	Wey-burn	The US Dakota Gasification Company operates a coal gasification unit that captures 3 million tonnes of CO_2 per year, and Canada's SaskPower Boundary Dam Power Station has a post-combustion capture unit handling 1 million tonnes annually.	320 and 80 km pipelines from the US to Canada and Saka power station to the oil field, respectively.	CO_2 storage capacity with EOR of the order of 1.8 $Mt.a^{-1}$ at a depth of 1450 m. The inception of injection took place in Oct 2000.	So far, approximately 35 million tonnes of CO_2 have been stored.	This is among the world's biggest projects for CCS, with the most advanced and thorough monitoring systems. Its monitoring includes multiple rounds of 3D seismic surveys with different components, detailed seismic monitoring inside wells, passive seismic tracking, and surface environmental checks. It has collected the largest geological CO_2 storage datasets ever and served as the basis for developing the CSA Z741 standard for CO_2 storage in Canada and the US.
02	Aquis-tore	Canada's SaskPower Boundary Dam Power Station operates a post-combustion capture unit that captures 1 million tonnes of CO_2 per year.	A 10-km pipeline (covering 3.4 km in a straight line) transports CO_2 to the saline aquifer storage site.	The storage is being done underground at a depth of 3400 m in the saline aquifer. The injection process started in 2015.	To date, approximately 3.5 lakh tonnes of CO_2 have been stored.	This is the world's deepest geological CO_2 storage project, located 3,400 metres underground, and it uses some of the most advanced and challenging monitoring technologies. The site is equipped with permanent 3D seismic monitoring, borehole fibre optic DAS VSP, deep-well fibre optic temperature and pressure sensors, passive seismic monitoring, tiltmeters, and GPS for tracking surface movements, environmental monitoring, and more. Building on lessons from the Weyburn Project, it helped develop permanent monitoring tools and techniques for CO_2 storage. To ensure the monitoring system's reliability and to study natural background conditions, multiple rounds of 3D seismic and environmental monitoring were done before CO_2 injection began.

03	Sleipner (Norway)	At the Sleipner Vest Field, about 850,000 tonnes of CO_2 are captured each year from natural gas using chemical solvents through an amine absorption process.	CO_2 is separated on an offshore platform and then injected deep into saline aquifers beneath the seafloor.	In this, the storage is being done deep in the saline aquifers at a depth of 800–1100 m. This is one of the first in nature, as per the offshore site considerations, with the injection process started in 1996.	To date, the stored amount of CO_2 is approximately 17 million tonnes.	CO_2 is injected into two separate saline sandstone aquifers. The first time-lapse world project depicts that the reservoir's density gets reduced after CO_2 injection. The site underwent eight rounds of 3D seismic monitoring using marine streamers, along with advanced deep subsurface monitoring that was both technically complex and sophisticated. The main goal was to understand how fluids move through the reservoir after CO_2 storage underground. Using comprehensive monitoring techniques, researchers collected detailed data that helped assess the environment and ensure the CO_2 is stored safely in these complex reservoirs.
04	CO_2 CRC Otway, Australia	CO_2 is extracted from natural gas, which contains 80% CO_2 and 20% methane.	Established pipeline system of 2.25 km.	Underground saline aquifers at a depth of 1565 m. Injection was initiated in the year 2009.	So far, an approximate total of 80,000 tonnes of CO_2 has been stored.	This project assesses the injection rate at different stages to meet scientific objectives. It explores imaging and geophysical monitoring technologies across various CO_2 injection rates, especially at low rates. It is the first facility in the world to study how CO_2 injection affects a fault's ability to seal. The site boasts the most comprehensive experimental facilities available, both on-site and indoors, with equipment for monitoring surface deformation, the environment, borehole fibre optic DAS VSP, passive seismic activity, permanent 3D seismic, and much more.
05	Ketzin Germany	Schwarze Pumpe Capture is a small-scale project focused on industrial hydrogen production.	Instead of pipelines, vehicular transportation is being used.	Underground storage in saline aquifers at a depth of 630–650 m. The injection process started in the year 2008.	67,271 t of CO_2 have been stored.	This project uses the world's most efficient 4D seismic monitoring and the most effective resistivity techniques, combining both surface and borehole geophysical monitoring. Long-term monitoring continued even after the wells were closed. The use of comprehensive monitoring technology ensured the safe storage of CO_2 in complex reservoirs and helped assess the surrounding environment. The analysis achieved a highly accurate prediction of CO_2 reserves and storage capacity. The main research focus was on understanding how fluids move through the reservoir after CO_2 is stored underground.

S.N	Project name	Source and scale	Transportation	Storage System	Approx. Amount of CO_2 Stored	Characteristics features
06	Tomakomai, Japan	High-concentration CO_2 is captured at a rate of 100,000 tonnes per year during the hydrogen generation process at processing units, through industrial separation and chemical adsorption methods.	CO_2 is injected from the capture point into a land-based injection well.	Stored in two seabed saline aquifers, one at 1,000 metres deep and the other at 3,000 metres deep, using a horizontal well injection method.	Approximately, so far, a total of 300,110 tonnes of CO_2 have been stored.	A chemical absorption process was used with newly developed absorbing solvents that have excellent resistance to thermal degradation and exceptional CO_2 absorption performance. This process achieved efficient CO_2 liquid–vapour separation while conserving energy at high pressures (3–5 atm). The HiPACT equipment, co-developed by JGC and BASF, was used to capture the CO_2, significantly lowering the energy and cost burdens of CCS projects. The cost of CO_2 recovery and compression was reduced by 25% to 35%. CO_2 was injected from land into two seabed formations: one made of sandstone and the other basalt. A unique ocean-bottom cable was developed, and sophisticated monitoring systems were used, including four-component time-lapse seismic, 4D marine streamer seismic, ocean-bottom seismometers, land borehole monitoring, and earthquake network-combined monitoring.
07	In Salah, Algeria	In the BP Salah Oil Field in Algeria, CO_2 with a 5.5% concentration was separated from natural gas at a rate of 1.0 to 1.2 million tonnes of CO_2 per year.	CO_2 is separated directly from natural gas at the wellhead and then transported to the CO_2 injection well.	Stored in a deep saline aquifer, which is 20 metres thick and located 1,880 metres underground. Injection started in August 2004 and was stopped in June 2011 due to concerns about potential caprock leakage, as indicated by monitoring data.	A total of 3.8 million tonnes of CO_2 were stored.	CO_2 is injected into a single set of saline aquifers. The project conducted the first-ever surface deformation monitoring study using InSAR, which revealed significant surface deformation after CO_2 was injected deep into the saline aquifer. This deformation could increase the risk of leakage through the caprock. To monitor the site, tiltmeter/GPS, 3D seismic monitoring (repeated several times), and micro-seismic monitoring were used. During wellbore integrity checks, leakage was found in the borehole wall, possibly caused by a geochemical reaction between the CO_2 and well cement.

1.8 Concluding Remarks

Looking into the boost of technology in both developed and developing countries, CCS efforts will eventually progress beyond capturing high-concentration CO_2 from sources like coal chemical plants, natural gas associated with flooding and storage, along with storage in saline aquifers, to low-concentration CO_2 capturing. For capturing low-concentration CO_2 tail gases from coal-fired power plants, refineries, hydrogen production, steelworks, cement factories, and similar industries will act as the primary sources. This shift is based on the experience gained by developed countries in CCS development. While the main challenge for geological storage is to carry out effective field-level monitoring, measurement, and verification to confirm storage capacity, emission reductions, and safety, the key challenge for capture technology lies in efficiently capturing large volumes of low-concentration CO_2 tail gases with minimal energy use. Building CCS infrastructure in developed countries is crucial for tackling key scientific challenges, reducing the overall cost of CCS technology, ensuring the safety of long-term underground storage, and driving innovation. It also supports commercialisation, promotion, demonstration, and training of experts. However, because geological conditions vary widely, it's often not possible to directly apply technologies from one place to another and get the same results. That's why sharing research findings and expertise from advanced countries is so important: it helps other nations in reducing costs, risk factors, and thereby successfully develop their own CCS projects. However, the large-scale growth of CO_2-EOR and storage has been slowed by limited research and understanding of key scientific issues like how CO_2 moves and accumulates underground, its storage status, and safety. Even if many large CCS projects were built, managing them all at once would be impossible without support from scientific research institutions. Without this guidance, it would be difficult to achieve rapid, large-scale emission reductions.

CCS hubs are seen as the future of commercial CCS projects because they help lower costs and reduce large-scale emissions from multiple sources in a region. These hubs bring together various carbon emitters with nearby suitable storage sites and establish efficient carbon transport networks. However, storage sites used in CCS hubs need to meet higher standards, so selecting the right geological locations carefully and precisely is essential. The size of the pipelines and the overall scale of the CCS hub depend on the storage site's ability to inject and hold CO_2. The main path for future growth in geological CO_2 storage is to permanently store CO_2 in saline aquifers, followed by storage in oil reservoirs and retired gas fields. Because saline aquifers can be located near a wider variety of high-carbon emission sources, they help reduce the need for long-distance CO_2 pipelines. This not only cuts costs but also lowers the carbon footprint associated with manufacturing and installing steel pipes. Additionally, storing CO_2 in saline aquifers doesn't require the CO_2 to be purified to a very high level, which further reduces collection costs. Implementing CCS requires improving energy efficiency and reducing the carbon footprint at every stage, looking at the whole process to achieve the biggest net emission reductions at the lowest possible cost.

Finally, the large-scale CCS projects in developed countries typically take 5–10 years to complete, from choosing the site to full operation. According to the IEA and CSLF, meeting the Paris Agreement's long-term temperature goals won't be possible without significant CCS emission reductions by 2030. Incentives like carbon taxes and regulations that limit emissions play a key role in driving CCS growth in countries like the US, Norway, and Canada. Meanwhile, carbon markets in China and the European Union can help promote CCS technology. As carbon prices increase, CCS will gradually become more profitable and commercially viable.

REFERENCES

[1] N. Stern, The economics of climate change, in: Am. Econ. Rev., 2008: pp. 1–37. https://doi.org/10.1257/aer.98.2.1.

[2] Intergovernmental Panel on Climate (IPCC) 2018, IPCC, 2018: Global Warming of 1.5°C. An IPCC Special Report on the impacts of global warming of 1.5°C above pre-industrial levels and related global greenhouse gas emission pathways, in the context of strengthening the global response to the threat of cli, Spec. Rep. Glob. Warm. 1.5°C 2 (2018) 1–631. https://cir.nii.ac.jp/crid/1130574323982454028.

[3] M.Z. Jacobson, Review of solutions to global warming, air pollution, and energy security, Energy Environ. Sci. 2 (2009) 148–173. https://doi.org/10.1039/b809990c.

[4] McKinsey, Carbon Capture & Storage: Assessing the Economics, Carbon N. Y. (2008). http://www.mckinsey.com/clientservice/ccsi/pdf/CCS_Assessing_the_Economics.pdf.

[5] G.T. Rochelle, Amine Scrubbing for CO_2 Capture, Science (80). 325 (2009) 1652–1654. https://doi.org/10.1126/science.1176731.

[6] Intergovernmental Panel on Climate Change, IPCC Special Report on Carbon Dioxide Capture and Storage, 2011.

[7] Madhusudhana M. Devadiga et al., Describing critical barriers in commercial deployment of CO2 capture and utilization technologies, Energy Environ. Sci., 2026,19, 743-790https://doi.org/10.1039/D6EE00134C

[8] R. Stuart Haszeldine, Carbon capture and storage: how green can black be?, Science (80). 325 (2009) 1647–1652. https://doi.org/10.1126/science.1172246.

[9] R. Haszeldine, Carbon sequestration, Science (80). 325 (2009) 815–830. https://doi.org/10.1098/rstb.2007.2185.

[10] T.A. Saleh, Nanomaterials and hybrid nanocomposites for CO_2 capture and utilization: environmental and energy sustainability, RSC Adv. 12 (2022) 23869. https://doi.org/10.1039/D2RA03242B.

[11] D.M. D'Alessandro, B. Smit, J.R. Long, Carbon dioxide capture: Prospects for new materials, Angew. Chemie - Int. Ed. 49 (2010) 6058–6082. https://doi.org/10.1002/anie.201000431.

[12] Jagdeep Singh et al., Tailoring carbon nanomaterial architectures for CO2 capture: structure–property relationships, surface engineering, and future perspectives, Materials Advances, 7, 6,2026,3031-3071, https://doi.org/10.1039/d5ma01408e

[13] P. Chowdhury, C. Bikkina, S. Gumma, Gas adsorption properties of the chromium-based metal organic framework MIL-101, J. Phys. Chem. C 113 (2009) 6616–6621. https://doi.org/10.1021/jp811418r.

[14] R. Krishna, Describing the diffusion of guest molecules inside porous structures, J. Phys. Chem. C 113 (2009) 19756–19781. https://doi.org/10.1021/jp906879d.

[15] D.W. Keith, M. Ha-Duong, J.K. Stolaroff, Climate strategy with CO2 capture from the air, Clim. Change 74 (2006) 17–45. https://doi.org/10.1007/s10584-005-9026-x.

[16] W.J. Ripple, C. Wolf, T.M. Newsome, P. Barnard, W.R. Moomaw, World Scientists' Warning of a Climate Emergency, Bioscience 70 (2020) 8–12. https://doi.org/10.1093/biosci/biz088.

[17] G. Supran, S. Rahmstorf, N. Oreskes, Assessing ExxonMobil's global warming projections, Science (80). 379 (2023). https://doi.org/10.1126/SCIENCE.ABK0063/SUPPL_FILE/SCIENCE.ABK0063_SM.PDF.

[18] 1949- Guterres, António, U. Secretary-General, Following three hottest days on record, Secretary-General launches global action call to care for most vulnerable, protect workers, boost resilience using data, science, (2024).

[19] UNEP, United Nations Environment Programme (2023). Emissions Gap Report 2023: Broken Record – Temperatures hit new highs, yet world fails to cut emissions (again). Nairobi. https://doi.org/10.59117/20.500.11822/43922., 2023. https://scholar.google.com/scholar?hl=en&as_sdt=0%2C5&q=+Emissions+Gap+Report+2023%3A+Broken+Record%3A+Temperatures+Hit+New+Highs%2C+yet+World+Fails+to+Cut+Emissions+%28Again%29+.+UNEP&btnG=.

[20] B. Hönisch, D.L. Royer, D.O. Breecker, P.J. Polissar, G.J. Bowen, M.J. Henehan, Y. Cui, M. Steinthorsdottir, J.C. McElwain, M.J. Kohn, A. Pearson, S.R. Phelps, K.T. Uno, A. Ridgwell, E. Anagnostou, J. Austermann, M.P.S. Badger, R.S. Barclay, P.K. Bijl, T.B. Chalk, C.R. Scotese, E. de la Vega, R.M. DeConto, K.A. Dyez, V. Ferrini, P.J. Franks, C.F. Giulivi, M. Gutjahr, D.T. Harper, L.L. Haynes, M. Huber, K.E. Snell, B.A. Keisling, W. Konrad, T.K. Lowenstein, A. Malinverno, M. Guillermic, L.M. Mejía, J.N. Milligan, J.J. Morton, L. Nordt, R. Whiteford, A. Roth-Nebelsick, J.K.C. Rugenstein, M.F. Schaller, N.D. Sheldon, S. Sosdian, E.B. Wilkes, C.R. Witkowski, Y.G. Zhang, L. Anderson, D.J. Beerling, C. Bolton, T.E. Cerling, J.M. Cotton, J. Da, D.D. Ekart, G.L. Foster, D.R. Greenwood, E.G. Hyland, E.A. Jagniecki, J.P. Jasper, J.B. Kowalczyk, L. Kunzmann, W.M. Kürschner, C.E. Lawrence, C.H. Lear, M.A. Martínez-Botí, D.P. Maxbauer, P. Montagna, B.D. A. Naafs, J.W.B. Rae, M. Raitzsch, G.J. Retallack, S.J. Ring, O. Seki, J. Sepúlveda, A. Sinha, T.F. Tesfamichael, A. Tripati, J. van der Burgh, J. Yu, J.C. Zachos, L. Zhang, Toward a Cenozoic history of atmospheric CO_2, Science (80). 382 (2023). https://doi.org/10.1126/SCIENCE.ADI5177/SUPPL_FILE/SCIENCE.ADI5177_SM.PDF.

[21] L. Cheng, J. Abraham, K.E. Trenberth, T. Boyer, M.E. Mann, J. Zhu, F. Wang, F. Yu, R. Locarnini, J. Fasullo, F. Zheng, Y. Li, B. Zhang, L. Wan, X. Chen, D. Wang, L. Feng, X. Song, Y. Liu, F. Reseghetti, S. Simoncelli, V. Gouretski, G. Chen, A. Mishonov, J. Reagan, K. Von Schuckmann, Y. Pan, Z. Tan, Y. Zhu, W. Wei, G. Li, Q. Ren, L. Cao, Y. Lu, New Record Ocean Temperatures and Related Climate Indicators in 2023, Adv. Atmos. Sci. 41 (2024) 1068–1082. https://doi.org/10.1007/S00376-024-3378-5/METRICS.

[22] J. Esper, M. Torbenson, U. Büntgen, 2023 summer warmth unparalleled over the past 2,000 years, Nature 631 (2024) 94–97. https://doi.org/10.1038/S41586-024-07512-Y;TECHMETA=141;SUBJMETA=106,413,694,704;KWRD=CLIMATE+CHANGE,PALAEOCLIMATE.

[23] W.J. Ripple, C. Wolf, J.W. Gregg, J. Rockström, T.M. Newsome, B.E. Law, L. Marques, T.M. Lenton, C. Xu, S. Huq, L. Simons, D. Anthony King, The 2023 state of the climate report: Entering uncharted territory, Bioscience 73 (2023) 841–850. https://doi.org/10.1093/BIOSCI/BIAD080.

[24] M. National Academies of Sciences, Engineering, Reflecting Sunlight: Recommendations for Solar Geoengineering Research and Research Governance, 2021. https://scholar.google.com/scholar?hl=en&as_sdt=0%2C5&q=Recommendations+for+Solar+Geoengineering+Research+and+Research+Governance&btnG=.

[25] W.J. Ripple, C. Wolf, J.W. Gregg, J. Rockström, M.E. Mann, N. Oreskes, T.M. Lenton, S. Rahmstorf, T.M. Newsome, C. Xu, J.C. Svenning, C.C. Pereira, B.E. Law, T.W. Crowther, The 2024 state of the climate report: Perilous times on planet Earth, Bioscience 74 (2024) 812–824. https://doi.org/10.1093/biosci/biae087.

[26] A. Robinson, J. Lehmann, D. Barriopedro, S. Rahmstorf, D. Coumou, Increasing heat and rainfall extremes now far outside the historical climate, Npj Clim. Atmos. Sci. 4 (2021) 1–4. https://doi.org/10.1038/S41612-021-00202-W;SUBJMETA=106,172,704;KWRD=CLIMATE+SCIENCES,ENVIRONMENTAL+SCIENCES.

[27] K.L. Ebi, A. Capon, P. Berry, C. Broderick, R. de Dear, G. Havenith, Y. Honda, R.S. Kovats, W. Ma, A. Malik, N.B. Morris, L. Nybo, S.I. Seneviratne, J. Vanos, O. Jay, Hot weather and heat extremes: health risks, Lancet 398 (2021) 698–708. https://doi.org/10.1016/S0140-6736(21)01208-3/ASSET/AC2E3883-2ECF-49E4 AEBF70A724063768/MAIN.ASSETS/GR1.JPG.

[28] J.T. Howard, N. Androne, K.C. Alcover, A.R. Santos-Lozada, Trends of Heat-Related Deaths in the US, 1999-2023, JAMA 332 (2024) 1203–1204. https://doi.org/10.1001/JAMA.2024.16386.

[29] C. WH, Extreme Weather: And What to Do., CO_2 Found. (2020). https://scholar.google.com/scholar?q=-Calvin+WH+.+2020.+Extreme+Weather:+And+What+to+Do+about+It.+CO2+Foundation.+https://www.amazon.com/EXTREME-WEATHER-+what-do-about-ebook/dp/B0841R8LV7&hl=en&as_sdt=0,5.

[30] A.J. Woodhead, C.C. Hicks, A. V. Norström, G.J. Williams, N.A.J. Graham, Coral reef ecosystem services in the Anthropocene, Funct. Ecol. 33 (2019) 1023–1034. https://doi.org/10.1111/1365-2435.13331;PAGE:STRING:ARTICLE/CHAPTER.

[31] O. Hoegh-Guldberg, E.S. Poloczanska, W. Skirving, S. Dove, Coral reef ecosystems under climate change and ocean acidification, Front. Mar. Sci. 4 (2017) 252954. https://doi.org/10.3389/FMARS.2017.00158/XML/NLM.

[32] T.P. Hughes, J.T. Kerry, A.H. Baird, S.R. Connolly, A. Dietzel, C.M. Eakin, S.F. Heron, A.S. Hoey, M.O. Hoogenboom, G. Liu, M.J. McWilliam, R.J. Pears, M.S. Pratchett, W.J. Skirving, J.S. Stella, G. Torda, Global warming transforms coral reef assemblages, Nature 556 (2018) 492–496. https://doi.org/10.1038/S41586-018-0041-2;TECHMETA=120,129;SUBJMETA=158,2165,631,704,826,829;KWRD=CLIMATE-CHANGE+ECOLOGY,MARINE+BIOLOGY.

[33] Climate.gov (Organization), How does 2023-24 global coral bleaching compare to past events? | NOAA Climate.gov, (2024). https://www.climate.gov/news-features/featured-images/how-does-2023-24-global-coral-bleaching-compare-past-events.

[34] J.A. O'Donnell, M.P. Carey, J.C. Koch, C. Baughman, K. Hill, C.E. Zimmerman, P.F. Sullivan, R. Dial, T. Lyons, D.J. Cooper, B.A. Poulin, Metal mobilization from thawing permafrost to aquatic ecosystems is driving rusting of Arctic streams, Commun. Earth Environ. 5 (2024) 1–10. https://doi.org/10.1038/s43247-024-01446-z.

[35] D.W. Keith, Geoengineering the Climate: History and Prospect 1, in: Ethics Nanotechnology, Geoengin. Clean Energy, Routledge, 2020: pp. 207–246. https://doi.org/10.4324/9781003075028-15.

[36] D. Visioni, E. Slessarev, D.G. Macmartin, N.M. Mahowald, C.L. Goodale, L. Xia, What goes up must come down: Impacts of deposition in a sulfate geoengineering scenario, Environ. Res. Lett. 15 (2020) 094063. https://doi.org/10.1088/1748-9326/ab94eb.

[37] K.P. Whyte, Indigeneity in Geoengineering Discourses: Some Considerations, in: Ethics "Geoengineering" Glob. Clim., Routledge, 2020: pp. 60–78. https://doi.org/10.4324/9781003049012-4.

[38] W. Smith, U. Bhattarai, D.G. MacMartin, W.R. Lee, D. Visioni, B. Kravitz, C. V. Rice, A subpolar-focused stratospheric aerosol injection deployment scenario, Environ. Res. Commun. 4 (2022) 095009. https://doi.org/10.1088/2515-7620/ac8cd3.

[39] B.K. Sovacool, C.M. Baum, S. Low, Determining our climate policy future: expert opinions about negative emissions and solar radiation management pathways, Mitig. Adapt. Strateg. Glob. Chang. 27 (2022) 1–50. https://doi.org/10.1007/S11027-022-10030-9/TABLES/1.

[40] J.D. Figueroa, T. Fout, S. Plasynski, H. McIlvried, R.D. Srivastava, Advances in CO_2 capture technology-The U.S. Department of Energy's Carbon Sequestration Program, Int. J. Greenh. Gas Control 2 (2008) 9–20. https://doi.org/10.1016/S1750-5836(07)00094-1.

[41] P.H.M. Feron, C.A. Hendriks, CO_2 capture process principles and costs, Oil Gas Sci. Technol. 60 (2005) 451–459. https://doi.org/10.2516/ogst:2005027.

[42] L.I. Eide, D.W. Bailey, Precombustion decarbonisation processes, Oil Gas Sci. Technol. 60 (2005) 475–484. https://doi.org/10.2516/ogst:2005029.

[43] Carbon capture – CSS - Vattenfall,. https://group.vattenfall.com/sustainability/climate-transition-plan/carbon-capture (accessed May 25, 2025).

[44] P.N. Dyer, R.E. Richards, S.L. Russek, D.M. Taylor, Ion transport membrane technology for oxygen separation and syngas production, Solid State Ionics 134 (2000) 21–33. https://doi.org/10.1016/S0167-2738(00)00710-4.

[45] M. Abu-Khader, Recent progress in CO_2 capture/sequestration: A review, Energy Sources, Part A Recover. Util. Environ. Eff. 28 (2006) 1261–1279. https://doi.org/10.1080/009083190933825.

[46] T.E. Gill, TEN YEARS OF HANDLING CO//2 FOR SACROC UNIT., in: Soc. Pet. Eng. AIME, SPE, OnePetro, 1982. https://doi.org/10.2118/11162-ms.

[47] J. Ma, L. Li, H. Wang, Y. Du, J. Ma, X. Zhang, Z. Wang, Carbon Capture and Storage: History and the Road Ahead, Engineering 14 (2022) 33–43. https://doi.org/10.1016/j.eng.2021.11.024.

[48] C. Marchetti, On geoengineering and the CO_2 problem, Clim. Change 1 (1977) 59–68. https://doi.org/10.1007/BF00162777.

[49] F.M. Orr, Onshore geologic storage of CO_2, Science (80). 325 (2009) 1656–1658. https://doi.org/10.1126/science.1175677.

[50] S. Chu, Carbon capture and sequestration, Science (80). 325 (2009) 1599. https://doi.org/10.1126/science.1181637.

[51] A. Chadwick, G. Williams, N. Delepine, V. Clochard, K. Labat, S. Sturton, M.L. Buddensiek, M. Dillen, M. Nickel, A.L. Lima, R. Arts, F. Neele, G. Rossi, Quantitative analysis of time-lapse seismic monitoring data at the Sleipner CO_2 storage operation, Lead. Edge (Tulsa, OK) 29 (2010) 170–177. https://doi.org/10.1190/1.3304820.

[52] B. Hitchon, Best practices for validating CO2 geological storage: Observations and guidance from the IEA GHG Weyburn-Midale CO2 Monitoring Project, Geosci. Publ. Sherwood Park (2012).

[53] K. Brown, S. Whittaker, M. Wilson, W. Srisang, H. Smithson, P. Tontiwachwuthikul, The history and development of the IEA GHG Weyburn-Midale CO_2 Monitoring and Storage Project in Saskatchewan, Canada (the world largest CO_2 for EOR and CCS program), Petroleum 3 (2017) 3–9. https://doi.org/10.1016/j.petlm.2016.12.002.

[54] GLOBAL CCS INSTITUTE, Global Status of CCS Report: 2020 - Global CCS Institute, (2020). https://www.globalccsinstitute.com/resources/publications-reports-research/global-status-of-ccs-report-2020/.

[55] S.R. Gislason, E.H. Oelkers, Carbon storage in basalt, Science (80). 344 (2014) 373–374. https://doi.org/10.1126/science.1250828.

[56] J.M. Matter, M. Stute, S. Snæbjörnsdottir, E.H. Oelkers, S.R. Gislason, E.S. Aradottir, B. Sigfusson, I. Gunnarsson, H. Sigurdardottir, E. Gunnlaugsson, G. Axelsson, H.A. Alfredsson, D. Wolff-Boenisch, K. Mesfin, D.F.D.L.R. Taya, J. Hall, K. Dideriksen, W.S. Broecker, Rapid carbon mineralization for permanent disposal of anthropogenic carbon dioxide emissions, Science (80-.). 352 (2016) 1312–1314. https://doi.org/10.1126/science.aad8132.

[57] Y. Sawada, J. Tanaka, D. Tanase, T. Sasaki, C. Suzuki, Overall Review of Tomakomai CCS Demonstration Project ~Target of 300,000 tonnes CO_2 injection achieved, in: 15th Greenh. Gas Control Technol. Conf. 2021, GHGT 2021, International Energy Agency Greenhouse Gas, IEAGHG, 2021. https://doi.org/10.2139/ssrn.3812038.

[58] D. White, Seismic characterization and time-lapse imaging during seven years of CO_2 flood in the weyburn field, saskatchewan, canada, Int. J. Greenh. Gas Control 16 (2013) S78–S94. https://doi.org/10.1016/j.ijggc.2013.02.006.

[59] K. Worth, D. White, R. Chalaturnyk, J. Sorensen, C. Hawkes, B. Rostron, J. Johnson, A. Young, Aquistore project measurement, monitoring, and verification: From concept to CO_2 injection, in: Energy Procedia, Elsevier, 2014: pp. 3202–3208. https://doi.org/10.1016/j.egypro.2014.11.345.

[60] J.P. Verdon, J.M. Kendall, A.L. Stork, R.A. Chadwick, D.J. White, R.C. Bissell, A comparison of geomechanical effects at three "megatonne" CO_2 storage sites - Sleipner, Weyburn and in Salah, in: Sustain. Earth Sci. SES 2013 Technol. Sustain. Use Deep Sub-Surface, National Academy of Sciences, 2013: pp. E2762–E2771. https://doi.org/10.3997/2214-4609.20131607.

Chapter 2

ADVANCED MATERIALS INTERFACES FOR CO_2 CAPTURE

Abstract: This book chapter emphasises the importance of reducing and utilising CO_2 as a fuel. For the proper capturing and utilising of CO_2, different materials were employed from time to time. However, with technological development, new methods for material fabrication came into being, such as additive manufacturing and 3D printing technologies. So, in this chapter, discussions related to the role of amine scrubbing, carbon and its composites, membranes, and COFs and MOFs were discussed. The discussion provides an elaborate overview of the progress made by the materials so far, with the advent of technology. After an enlightening discussion, a brief description of the different challenges faced by these materials is also provided.

2.1 Introduction

Energy usage per capita is a key indicator when assessing the standard of living in a country. It plays an essential role in understanding how well a nation is meeting the needs of its population. Given the rising global population, managing energy resources sustainably has become a critical task for governments [1]. The goal is to meet energy demands while ensuring the well-being of both current and future generations. Looking ahead, it is highly expected that global energy consumption will grow by 48% over the next 28 years, rising from 549 quadrillion Btu in 2012 to 815 quadrillion Btu in 2040 [2]. As illustrated in **Figure 2.1 (a)**, it is evident that fossil fuels will remain the easily accessible source of energy for the foreseeable future. However, relying on fossil fuels for power generation will inevitably lead to a rise in CO_2 emissions unless strict measures are put in place.

Additionally, fossil fuel usage in transportation, industry, and agriculture contributes significantly to global CO_2 emissions, further accelerating the rise in atmospheric CO_2 levels. In fact, CO_2 is responsible for about 77% of all greenhouse gas emissions globally [3]. In response to these challenges, there is an increasing push to focus on renewable energy sources, such as solar, wind, and tidal power. While transitioning to renewable sources is vital, it's also worth considering ways to decrease the consumption of non-renewable sources of energy, like coal, oil, and natural gas, or even replace them entirely. Yet, factors such as geographic location, as well as seasonal and nighttime variations, influence the availability of renewable resources. Moreover, utilising solar and wind energy requires large expanses of land, which presents its own set of challenges.

2.2 Management of CO_2: Necessity and Technology

According to several calculations, we must maintain the concentration of greenhouse gases below 450 parts per million (ppm) of CO_2-equivalent in order to prevent the rise in the global temperature from exceeding 2°C at the end of the current century [4]. CO_2 levels have already surpassed 400 parts per million, which is perilously near that goal, according to readings taken from Mauna Loa in Hawaii. We need to reduce human-caused emissions by 40%–70% by 2050 compared to 2010 levels and eradicate them by 2100 in order to stay inside the 450-ppm limit [3]. Therefore, the current exercise of utilising fossil fuels as an energy source must end, with improved control of energy consumption, CO_2 emissions, and management is urgently needed for minimising CO_2 emissions and, therefore, global warming [5]. There have been several strategies put forth to lower CO_2 emissions worldwide, as highlighted below:

a) Enhancement of process efficiency to lower energy usage.

b) Using renewable and alternative energy sources like solar, wind, and geothermal, instead of the fossil fuels we rely on today.

c) Using technology to capture, store and make use of CO_2 (often called CCS & U).

To manage energy sustainably, we need to combine all three strategies mentioned earlier carefully. But beyond just performance, this approach must also address energy use, cost-effectiveness, and both direct and indirect CO_2 emissions. Finding the best options means closely looking at the accessible energy sources, the amount of CO_2 they produce, how they're managed, and how sustainable the manufacturing processes are. The natural way of maintaining the greenhouse gases with a primary focus on CO_2, through processes like photosynthesis and absorption by the oceans, is no longer enough to keep up with the increasing emissions from industries. Because of this, CO_2 levels have risen past the safe target of 450 ppm [6]. In this regard, different approaches have been employed in the CO_2 capturing technology and utilisation as discussed in **Chapter 1**. An outline of the research technologies explored to tackle the problem for capture and utilisation (CCU), along with the storage of CO_2 (CCS), is shown in **Figure 2.1 (b)**.

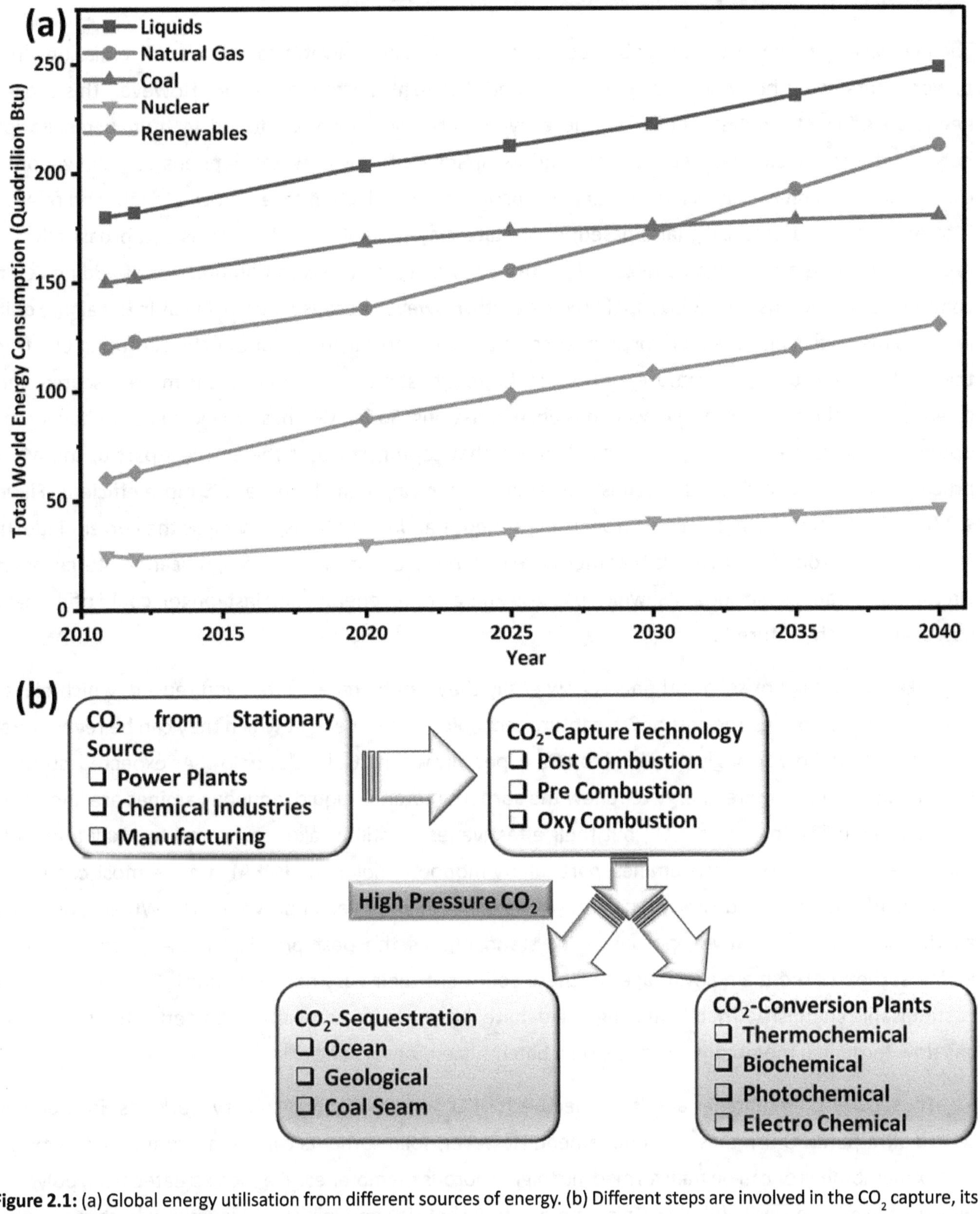

Figure 2.1: (a) Global energy utilisation from different sources of energy. (b) Different steps are involved in the CO_2 capture, its sequestration, and conversion process. Copyright Credit: (Ref. [2], U.S. Energy Information Administration, 2016 and Ref. [3], Chemical Engineering Research and Design, Elsevier, 2018)

2.3 CO_2 Capture Materials and Their Progress

The oceans are the largest natural absorbers of CO_2, and using seawater to capture CO_2 emissions from power plants could be an ideal solution, provided it doesn't harm marine life. However, this process needs an efficient catalyst, similar to the enzyme carbonic anhydrase found in human respiration, because the conversion between CO_2 gas and its dissolved forms in water happens very slowly on its own [7]. An alternative approach of creating an absorbent material that takes in CO_2 at night and releases it at dawn offers a promising land-based, solid-state way to capture carbon. This setup basically acts like a greenhouse around plants **(Figure 2.2 (a))**. It's inspired by the common practice of adding CO_2 to commercial greenhouses to help crops in their growth. However, because the CO_2 levels in the air are quite low, this method is better suited for smaller-scale uses. It's no surprise that directly capturing CO_2 from the air (known as direct air capture, or DAC) is the toughest method, and it needs more research since it tends to be the most expensive way to reduce emissions. So far, CCS has proved to be the favourable option for the zero-emission goal. It's well known that capturing CO_2 is the costliest part of the whole process, which is why it needs the most research and development to make CCS more efficient **(Figure 2.2 (b))**. The high costs mostly come from using aqueous amine solutions, as well as the current industry standard. The problem with this is that they have high heat requirements and significant material losses. On the other hand, solid sorbents, which use less energy to regenerate and last longer, could offer major cost savings in the future [8].

A key advantage of solid sorbents is how easily they can be regenerated and reused, which plays a major role in estimating the cost of CO_2 capture technology. The speed at which they can be regenerated is closely tied to how strongly they bind to CO_2, especially when CO_2 levels are low. As expected, stronger CO_2 binding requires more energy to renew the sorbent material. Liquid, aqueous amines are known for their strong ability to capture CO_2, but their effectiveness is still weaker compared to alkali and earth alkali metal carbonates. These amines, particularly monoethanolamine (MEA), are the most commonly used absorbents in the industry, though they are expensive and need improvement [9]. When considering all the potential alternatives, porous solids stand out as the best practical replacement for liquid amine sorbents. The main advantages of these solids are their easy handling, quick CO_2 diffusion, and customisable chemistry. There's still ongoing debate, however, about whether it's better to use a steady gas flow through a membrane or to rely on a batch system of sorbents [10].

Thus, current membranes aren't as selective for CO_2 as sorbents are, so they work best in situations where ultra-pure CO_2 isn't a strict requirement. However, solid sorbents can be incorporated into mixed-matrix membranes or can be transformed into highly porous membranes, like those created from polymers with built-in microporosity (PIM) [11]. This approach could help in combining the strengths of both membrane and sorbent technologies. A combined sorbent-membrane system could probably meet all the necessities for high-flux CO_2 required for wet streams [7]. Although solid adsorbents and catalysts show great potential, there are still challenges to overcome, such as slower reaction times, wear and tear over multiple cycles, and high-pressure drops. Traditional methods rely on the transformation of the powder into pellets or granules, but these often don't work and lack the efficiency as expected [12]. However, the extrusion-based monoliths help in reducing some of these issues, but they still face limitations when it comes to creating precise structures. These issues were effectively addressed through additive manufacturing, or 3D printing, which allows for specific regulator over channel dimensions, cell density, and wall thickness. Different researchers have explored the use of 3D printing in adsorption and catalysis, focusing on the principles of the printing process for designing the different materials [13]. Since 3D printing offers a promising way to improve the energy efficiency and material performance through precisely engineering their structures, it's gaining attention globally. This approach also helps to optimise key properties like adsorption capacity, reaction speed, selectivity, and conversion in adsorbents and catalysts.

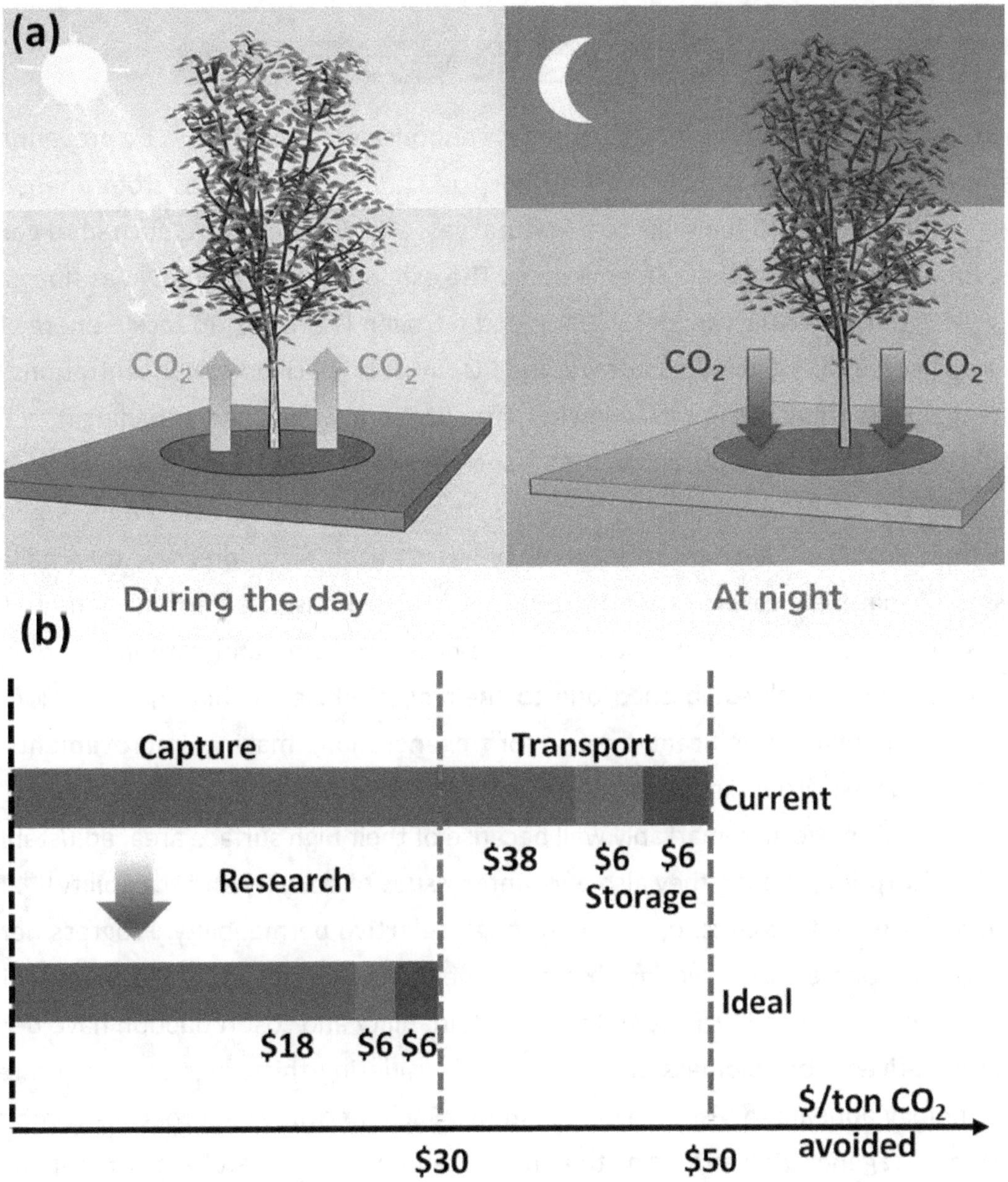

Figure 2.2: (a) Working mechanism of an ideal adsorbent promoting the growth of vegetation. (b) Approximate estimated cost for the carbon capture and CCS process. Copyright Credit: (Ref. [7], ChemSusChem, Wiley, 2017)

Additionally, 3D printing may reduce waste and maximise resource utilisation, supporting sustainability objectives. Additionally, the capability of recycling and reuse of printing materials, along with resin, supports the goals of the circular economy [14]. This book chapter will emphasis on the growth and progress of different materials employed for the CO_2 CCS. With an emphasis on adsorbents, catalysts, and membranes, this study thoroughly evaluates the developing area of 3D-printed materials in CO_2 collection, conversion, and separation. It includes an overview of 3D printing methods, a state-of-the-art analysis of 3D-printed membranes and adsorbents for CO_2 extraction from gas mixtures, and 3D-printed catalysts for a range of CO_2 conversion uses. By critically analysing the state-of-the-art materials available now, this study predicts that research in this dynamic and quickly changing field will continue to grow.

2.4 Carbon Dioxide Capture: Materials, Advancements, and Working

One of the most effective ways to reduce global greenhouse gas emissions is by preventing CO_2 from being released at the source. A major contributor to CO_2 emissions comes from energy production sources, where different fossil fuels like coal, natural gas, and petroleum are burned to generate heat, accounting for about 72% of global CO_2 emissions. The exhaust, primarily known as flue gas, accounts for about 7–15% of CO_2, water vapour (5–15%), and nitrogen (75–85%). To tackle energy inefficiency, especially in processes like geological storage, the CO_2 must be purified to concentrations higher than 90% [15]. The main challenge lies in extracting CO_2 from these mixtures and enriching it to the required purity level. Looking into the importance, in this section of the book chapter, we will try to enlighten the chemistry of some of the well-known materials, that is, absorbents, adsorbents, and membranes, along with their advanced version (3D-technology-based) used for addressing the challenge. These broad classes of materials cover a wide variety that has demonstrated strong potential for carbon capture. Absorbents include options such as amine-based solvents and carbonate solutions. These solvent systems remain well established due to their high efficiency, but they suffer from gradual degradation and demand significant energy for regeneration, making improvements necessary. Likewise, the advanced group known as adsorbents includes zeolites, activated carbon, and MOFs. These solid materials perform remarkably well because of their high surface area, adjustable pore size, and selective adsorption, though they also encounter issues of stability and scalability [16]. Membrane-based systems form another class, operating through selective permeability. Progress here has been marked by the development of mixed-matrix membranes containing nanoparticle fillers, which enhance both selectivity and permeability. Beyond these, sustainability and cost reduction have been achieved with sorbents derived from biomass and waste [17]. Considering this progress, the future of carbon capture technology appears to rest on hybrid systems. Such innovative approaches combine different classes with new regeneration strategies to minimise energy use. Careful assessments are essential, integrating techno-economic factors to ensure scalability, life-cycle benefits, and cost-effectiveness. To provide a broad perspective of these categories, each class will be discussed in detail in the following sections.

Amine Scrubbing: Employing the unique and efficient physical and chemical properties possessed by the CO_2 as compared to the nitrogen gas, absorbents play a vital role in the extraction of CO_2 from the flue gases. In this regard, the potential example is the amine scrubbing process, which uses amine solutions, like piperazine (PZ), methyl diethanolamine (MDEA), ethyl diethanolamine (EDEA), and monoethanolamine (MEA) as absorbents for CO_2 separation with a design depicted in **Figure 2.3 (a)** [18, 19]. Amines are effective at separating CO_2 because they react with CO_2, unlike nitrogen, which remains inert. Although amines are highly selective and can separate CO_2 in a single step, they are not commonly used due to the high-energy cost involved in regenerating them after the CO_2 has been absorbed. This energy cost is a significant drawback, despite the process being effective **(Figure 2.3 (b))**.

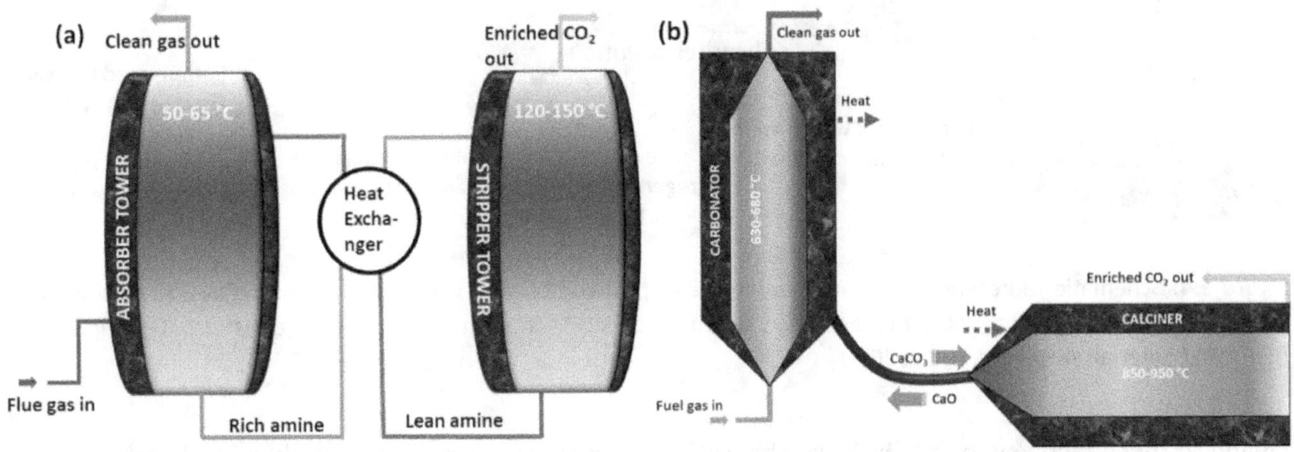

Figure 2.3: Schematic representation of the (a) amine and (b) lime, scrubbing processes.

Among the class of metal oxides, Magnesium oxide (MgO) and Calcium oxide (CaO) have shown promising results and have proved to be an effective chemical absorbent for CO_2 [20]. These oxides can absorb a significant amount of CO_2 through their reaction in the formation of bicarbonates and carbonates, with theoretical predicted absorption capacities of around 0.78g CO_2/g for CaO and 0.79g CO_2/g for MgO [21]. Besides that, these materials also show an excellent selectivity for CO_2 in a single step. Although being so efficient, there are some challenges with these oxide materials when used for CO_2 capture. Among the different challenges, some of the critical issues are: high regeneration temperatures (ranging from 600–900°C), substantial capacity loss over multiple cycles, and slow uptake kinetics. These issues render their capability for their ideal practical use [22].

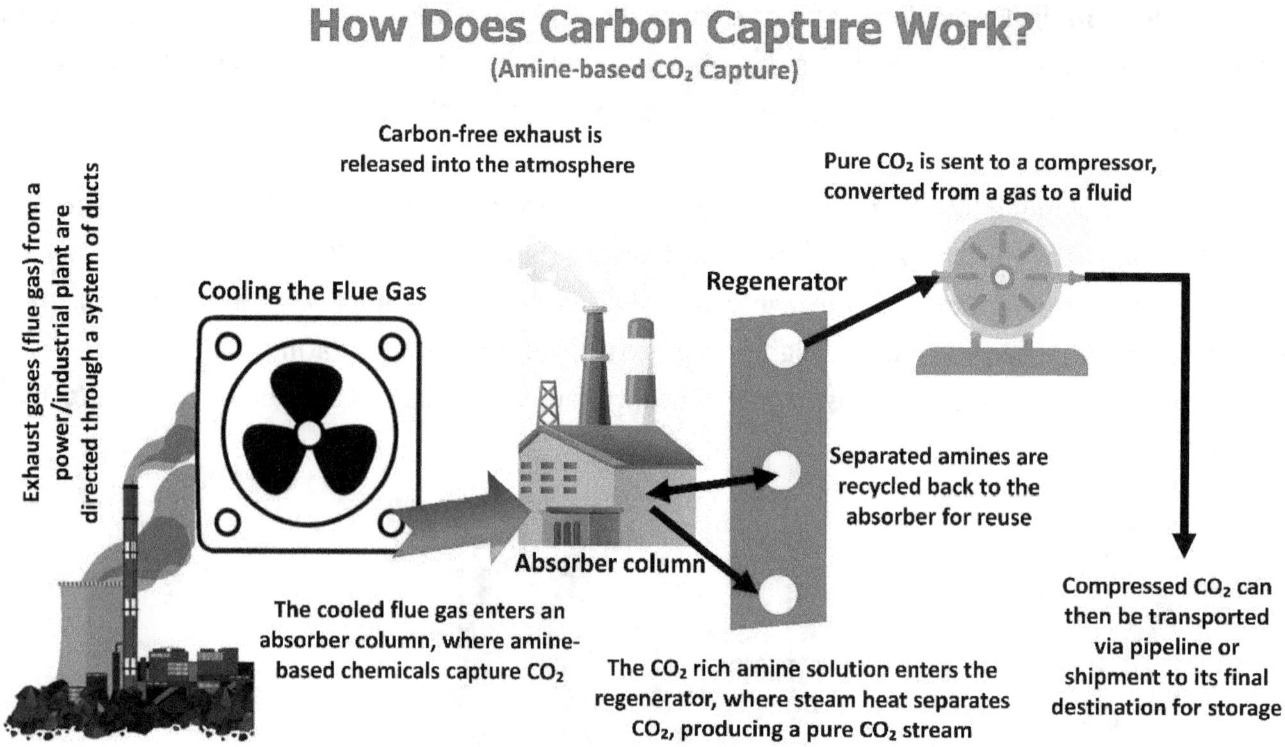

Figure 2.4: Schematic representation for the amine-based post-combustion CO_2 capturing, with flue gas-cooling, amine scrubber for CO_2 absorption, solvent regeneration system, recycling, with transport and storage systems. Copyright Credit: (Ref. [19], Materials Advances, RSC 2026)

Owing to these facts, even the chemical absorption techniques, such as lime washing and amines for CO_2 removal from emission sources, are still not very widely adopted/explored. The schematic representation of the amine-based CO_2 capturing units with complete setup is presented in **Figure 2.4**. Although with so many challenges a progress in using the amine as CO_2 capturing materials through 3D printing functionalisation is under development. In this regard, for enhancing CO_2 adsorption efficiency, amines are commonly used to modify popular adsorbents like silica, carbon, and MOFs [23]. The overview of the 3D technology is depicted in **Figure 2.5 (a-f)** [24]. In one innovative 3D-printing-based research study, monoliths made from multi-walled carbon nanotubes and polyethyleneimine (PEI) were printed using direct ink writing (DIW). The printed monoliths were then tested through Vacuum Swing Adsorption (VSA) in advance experiments, by employing a gas mixture of 15% CO_2 and 85% N_2. The aim was to achieve high levels of CO_2 purity (95%) and recovery (90%) while minimising energy consumption. During this process, the optimised cycle achieved a peak efficiency of 2.2 mol of CO_2 per m^3 of adsorbent per second (mol/m^3 ads s), with a minimum specific energy of 0.6 MJ/kg of CO_2 captured. While this study focused on process-level simulations to optimise cyclic adsorption with printed adsorbents, further research is needed to tackle challenges like amine leaching during cyclic operations and managing oxygen and moisture in real-world flue gas. A key hurdle lies in stabilising and optimising paste formulations, which are critical for scaling up to pilot plant operations. To address this, DIW was employed to print monoliths of silica powders, which were grafted with 3-aminopropyltrimethoxysilane (APS), polyethyleneimine (PEI), along with tetraethylenepentamine (TEPA) [25]. After five cycles, the APS-grafted monoliths demonstrated impressive cyclic stability, consistently maintaining their capacity. In contrast, PEI and TEPA-impregnated monoliths showed an enhancement in their adsorption capacity as the temperature rose, attributed to diffusional effects. On the other hand, the APS-grafted monoliths experienced a reduction in capacity with higher temperatures, which can be attributed to the minimal diffusional effects caused by the small APS monomers on the surface. This comparison underscores the adaptability of adsorbent functionalisation, printing methods, and dynamic behaviour across different conditions. According to a study by Sluijter *et al.* [26], the structure of printed silica, functionalised with amines, remained intact through a process of printing, debinding, and sintering. When compared to grafted silica pellets (0.93 mmol/g), it was observed that the amine-functionalised silica sorbents in 3D-printed monoliths exhibit a lower sorption capacity (0.66 mmol/g). The 3D sorbent supports applied, and the functionalised routes are depicted in **Figure 2.4 (g–i)**. Thus, the 3D-printed adsorbents offered clear advantages, such as a significantly reduced pressure drop (60 mbar/m vs 300 mbar/m) and improved mass transfer efficiency. These improvements suggest potential for further reductions in pressure drop and enhancements in capacity retention. Additionally, functionalising MCM-41 silica with amines possessing a large mesopore volume, could lead to competitive CO_2 adsorption capacity. Recently, structural topologies of MCM-41 functionalised with PEI were produced using DIW. In order to create the 3D-printable ceramic ink, the precursors employed as binder and plasticiser were methyl cellulose and bentonite, while MCM-41 was modified by varying the concentration (molecular weights) of PEI. It was found that the molecular weight of the PEI employed for functionalising MCM-41 influences the CO_2 adsorption capacity of the final 3D-printed monoliths across multiple temperatures. However, due to the inclusion of binders and plasticisers during the ink synthesis, the CO_2 absorption capacity of the 3D-printed monoliths was slightly lower in comparison to that of the powdered MCM-41 counterparts. The molecular weight of the PEI notably influenced the CO2 uptake capacity used to functionalise MCM-41, particularly at higher temperatures. As the molecular weight of PEI increases, the capacity was found to decrease. This was because the silica, held within the PEI matrix, restricts adsorption on the external surface of the aggregated adsorbent. However, since better dynamic efficiency can help mitigate the capacity loss caused by the binders, further investigation is required to understand the role of amine molecular weight variations in the process. Recent advancements have shown that combining amine-based ion-exchange resin with glass flakes coated in silver allows for the 3D printing of an electro-conducting sorbent, enhancing electrical conductivity [27]. It has been observed that sorbents containing 20 vol% silver flakes maintained strong electrical conductivity even at low humidity levels while being at room temperature. However, when humidity rises to 75%, the volume of the material usually expands by 7%, which raises resistance and reduces percolation. At silver flake concentrations above 25 vol%, percolation was achieved, enabling effective rejuvenation through Joule heating. Analysis demonstrated that these rejuvenation cycles could occur rapidly, with sorbent temperatures reaching up to 70°C within approximately 60 seconds of consistent heating. Sorbents developed through the printing method have shown a CO_2 adsorption capacity of the order of 1.5 mmol/g under 9 vol% CO_2. However, there was a slight decrease in capacity, around a 15–25% loss, depending on the proportion of sorbent powder in the resin [14]. This reduction in capacity was attributed to diffusion limitations caused by the silver particles. To improve temperature uniformity and minimise the risk of hot spots during cycling, adjustments are needed in the balance of conductive flakes, sorbent powder, adsorption, and percolation properties.

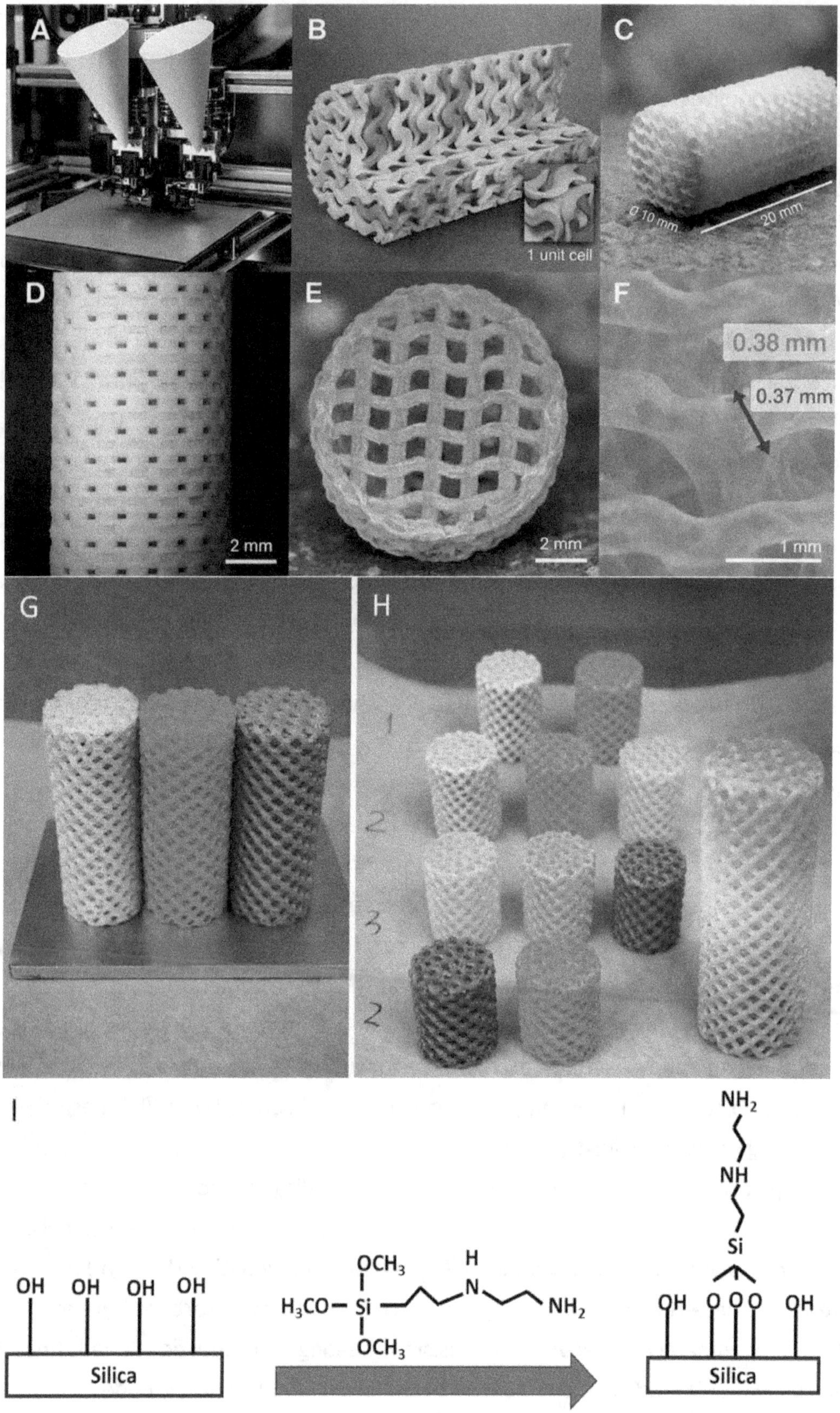

Figure 2.5: (a) MEX multi-material 3D printer, (b) Gyroid structure with 50% density developed using CAD design, (c–f) Microscopic images of SiO_2 of cylinders, (g, h) 3D printed SiO_2 sorbents, (i) Immo-Ammo functionalisation. Copyright Credit: (Ref. [24], Materials, MDPI, 2024. & Ref. [26], International Journal of Greenhouse Gas Control, Elsevier, 2021)

Carbon and its Composites: To achieve separation, porous adsorbents like zeolites and activated carbon leverage the differences in the physical properties of CO_2 and nitrogen (N_2). CO_2, with a smaller kinetic diameter of 3.30 Å, can diffuse more quickly than N_2, which has a kinetic diameter of 3.64 Å [28]. This allows CO_2 to enter the micropores (<2 nm) and mesopores (2-50 nm) that are predominant in these materials. Unlike amine scrubbing, which typically uses pressure swing adsorption (PSA) at normal temperature with vacuum assistance, the physical adsorption method allows for the recycling of activated carbon at lower temperatures. Activated carbon mainly relies on the pore size limits and its selectivity, thereby restricting its utility for multiple processes needed to achieve CO_2 purity above 90%, despite the fact that it requires low energy and is cost-efficient [29]. Besides that, these adsorbents exhibit slow kinetics and long equilibration times. Temperature also plays a significant role in their performance, as the uptake capacity tends to decrease as temperature rises. To optimise efficiency, flue gas must be cooled to around room temperature, which further reduces activated carbon's competitiveness. To make activated carbon more competitive with scrubbing methods, researchers have focused on enhancing its CO_2 selectivity, either by improving pore structures or developing composite materials [30].

The activated carbon materials can be enhanced in their activity through integration with chemical absorbents developed from metals and their oxides, such as MgO and CaO, to boost CO_2 selectivity through chemisorption. Some efforts have been made in merging activated carbon with various metals and their oxides to endorse chemisorption, thereby improving CO_2 selectivity as well as capturing capacity [30]. However, fillers made from porous materials like MOFs, zeolites, and silicates have also been used to enhance CO_2 sensitivity and optimise pore structure [30, 31]. While these techniques enhance the CO_2 selectivity of activated carbon, composite materials often face challenges, such for regeneration process, they require high temperature, possess high manufacturing costs, reduced porosity, and random filler arrangements. Additionally, a drawback of using CaO is that it is produced through the thermal decomposition of limestone, which results in the release of by-products in the form of CO_2 (in a stoichiometric amount). Water, a key constituent of the flue gas mixture, also faces some challenges. Amines, often used as aqueous solutions, benefit from water, which speeds up the scrubbing process. Studies show that water can improve the CO_2 absorption capabilities of lime-based oxides [32]. However, in these composite materials, water and CO_2 compete for the same binding sites, which can reduce efficiency [7]. Recently, the 3D version of the activated carbon has shown some promising improvements in enhancing its affinity for the CO_2 capture. Due to their strong affinity for CO_2, carbon-based adsorbents also show potential for use as 3D-printed materials. For instance, stereolithography (SLA) was employed to fabricate activated carbon monoliths, as depicted in **Figure 2.6 (a)** [33]. For evaluation of the performance, two monoliths featuring identical open-cell foam structures with variation in the degree of burn-off that occurred during the activation process, were tested in adsorption experiments carried out in a fixed-bed. These tests were carried out in a typical post-combustion process, with the temperature varying in the range of 313–373 K, at a partial CO_2 pressure of 120 kPa. Two monoliths were tested in the experiment, one with small burn-off and the other with larger burn-off, designated as M1 and M2. On comparative analysis, it has been observed that the monolith M1, which underwent a shorter burn-off time during activation, showed higher CO_2/N_2 selectivity, reaching a value of 18. Specifically, it achieved a selectivity ratio of 10 for a binary CO_2/N_2 mixture (15/85 vol%) at 313 K. This highlights how the activation duration of activated carbon monoliths plays an important role in both gas selectivity as well as enhancing the performance of thermal regeneration, thereby making it a key factor for effective use in temperature swing adsorption (TSA) processes. The direct 3D printing of porous carbon has also been explored, using an ink made from a combination of bentonite as the binder, poly(4-vinylphenol) gel, commercial activated carbon, and Pluronic F127 as a plasticiser as depicted in **Figure 2.6 (b)** [34]. The resulting carbon monolith was tested through 10 consecutive cycles of CO_2 adsorption followed by the desorption process, showing no significant loss or degradation in its ability to function as an adsorbent. Notably, this research marks the first thorough examination of creating 3D-printed porous carbon, irrespective of relying on the need for a sacrificial template. In such kinds of adsorbents, the commercial activated carbon serves as the primary source of porous carbon. At the same time, poly(4-vinylphenol) and Pluronic F127 are included for their dual roles as a plasticiser and as a contributor to meso-porosity. This material paves the way for further research into gas separation applications, showcasing the stability with effective adsorption capability of 3D-printed generated porous carbon. To assess the suitability of these monoliths for electric swing adsorption (ESA), DIW has been employed for printing the activated carbon monoliths. During the process potassium silicate has been employed as the binder as depicted in **Figure 2.6 (c)** [35].

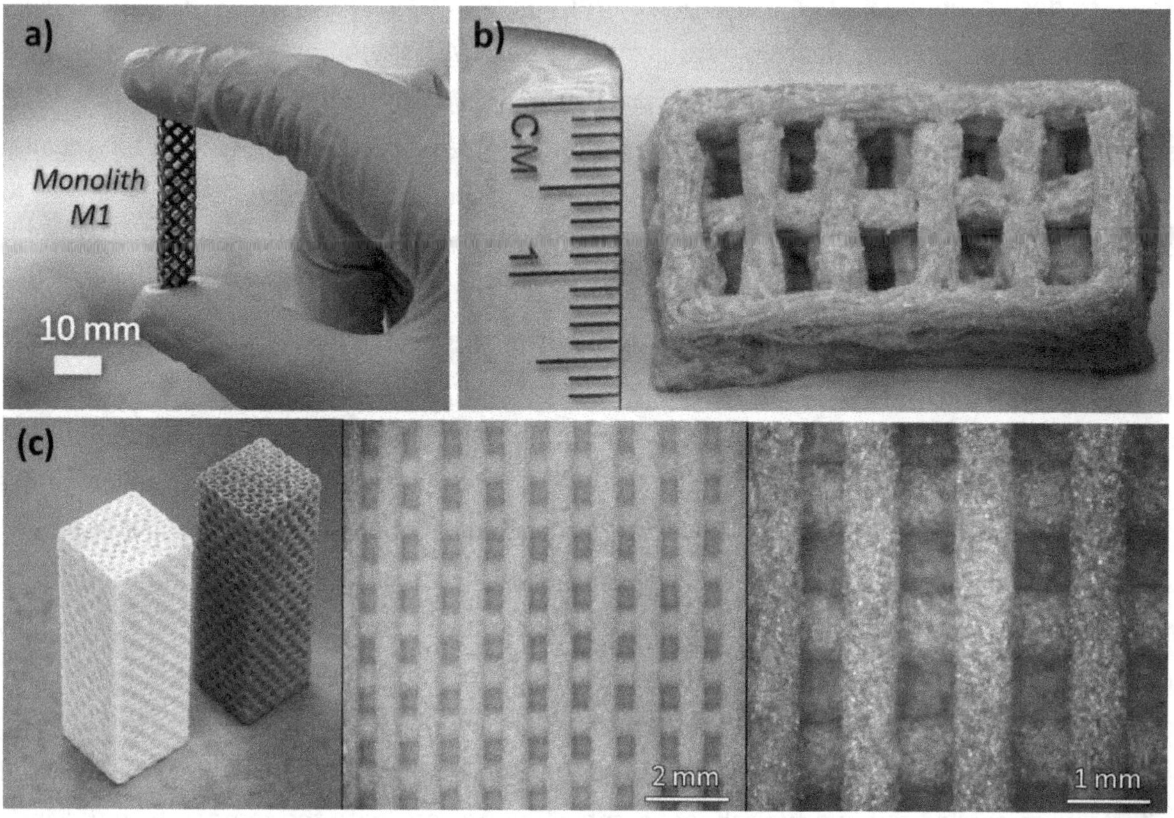

Figure 2.6: (a) Monolith under low temperature for physisorption process, (b) Image depicting porous and optical view of the 3D printed carbon monolith, (c) Images corresponding to the 3D printed carbon monolith and its magnified images after proper drying. Copyright Credit: (Ref. [33], Microporous & Mesoporous Materials, Elsevier, 2022, Ref. [34], Molecules, MDPI, 2022, and Ref [35], Separation & Purification Technology, Elsevier, 2022)

As a result, a shorter heating period led to less CO_2 being desorbed. Therefore, electrification played a key role in ensuring consistent heating across the adsorbent, reaching a temperature of 130°C in just 30 seconds with an 8 V potential. Leveraging the adsorbent's high conductivity presents a potential way to enhance heat transfer efficiency. However, to fully assess the effectiveness of electrical swing for quick rejuvenation, further research on a larger scale is needed, even though the results so far show promising potential. These results highlight that different choices of binders, printing techniques, and adsorbent materials result in collective progress in the mechanical strength, adsorption capacity, and dynamical effectiveness of the final monolithic structures. The comparative study not only sheds light on the trade-offs associated with various formulations but also opens up opportunities for making further progress in tailoring adsorbents for specific industrial CO_2 capture processes. The development of 3D-printed adsorbents for CO_2 collection can be significantly enhanced by exploring secondary growth methods, the effect of infusing the metal precursor, ESA, and optimising digital light processing [14].

COFs and MOFs: Developing porous materials with specific architectures for efficient CO_2 capture has proven to be a more promising approach. As a result, current research is focused on creating new porous materials that leverage variations in chemical properties for enhancing both kinetics and selectivity. In this context, different design strategies for porous organic polymers (POPs) have been explored, aiming to improve CO_2 selectivity and uptake capacity. The research underscores the significance of ultra-micropores (< 0.7 nm) and the incorporation of heteroatoms like nitrogen, which play a key role in enhancing selectivity through dipole-quadrupole interactions [29,36]. In this realm, COFs and MOFs have received remarkable attention in the field of CO_2 capture due to their marvellous properties, such as efficient porosity, tunable structures, lower structural density, and versatile pore sizes with strong thermal and chemical stabilities. The flexibility offered by the selection of linkers plays a crucial role in the design of COFs and MOFs, as it enables the creation of customised frameworks capable of absorbing CO_2 through physisorption and chemisorption [31,37]. COFs' ability to absorb CO_2 can be further enhanced by incorporating various functional groups, such as triazines, boronates, and imines, into their structure. For illustrations, it has been observed that the triazine-linked N_3-COF showed impressive CO_2 uptake capabilities of 19.8 and 11.8 wt.% at temperatures of 273 and 298 K, respectively, with pressure maintained at 1 bar, as shown in **Figure 2.7 (a)** [38]. Similarly, the imine-linked TpPa-COF, synthesised by Wang *et al.* [39], demonstrated an exceptional CO_2 uptake of 22.4 wt.% and a CO_2/N_2 selectivity of 32 at a temperature and pressure of 273 K and 1 bar, respectively. While COFs hold great potential as CO_2 adsorbents, their commercial use for isolating CO_2 from flue gas faces challenges related to scalability, repeatability, and crystallinity [40, 41].

In contrast, MOFs enhance CO_2 capture capabilities by modifying linkers, adjusting pore structures, and incorporating inorganic polynuclear aggregates, forming the clusters known as secondary building units. The CO_2 affinity of MOFs can be improved by adding metals like Mg, Zn, and Cu, along with nitrogen-containing moieties through the linkers [42]. For instance, Mg_2, synthesised by Long *et al.* [43], exhibited an impressive CO_2 capture capacity of 22.0 wt.% at 298 K at 1 bar pressure, with structure depicted in **Figure 2.7 (b)**. By further modifying Mg_2 using alkylamines, further enhancement in the efficiency of CO_2 capturing by 8.1 wt.% was achieved under atmospheric conditions of temperature 298 K and at pressure 0.39 mbar. The Cu MOF, PCN-88, developed by Li *et al.* [44] demonstrates a CO_2 capture capacity of 23.8 wt.% and 15.8 wt.% at a temperature of 273 K and 296 K, respectively, under 1 bar pressure. MOFs show great potential as adsorbents for both atmospheric CO_2 extraction and point-source CO_2 collection. In line with this potential, a Swiss company, NovoMOFAG, is working on commercialising MOFs for CO_2 collection applications.

Additionally, water, a crucial component of flue gas, must be considered in CO_2 capture systems, whether for atmospheric or flue gas CO_2 extraction [45]. It is well-established that some MOFs and COFs perform less effectively under wet conditions and may suffer from poor water stability. Besides that, the MOFs often struggle to retain their adsorption capabilities when subjected to the high temperatures during the calcination and sintering process, which are necessary for the removal of organic binders. This is primarily due to their poor thermal stability. To ensure the adsorption properties are preserved during 3D printing, it's crucial to use an appropriate inorganic binder in the right proportion with the MOF powder. Alternatively, MOF monoliths can be formed using organic binders like PVA through phase separation, which helps maintain the MOF's natural adsorption characteristics without the need for heat treatment to remove the binders. Additionally, the use of a solvothermal method for secondary MOF growth can effectively address issues related to binder presence and agglomeration [46].

In this regard, the MOF-74 monoliths **(Figure 2.7 (c))** were successfully 3D printed using DIW, where they were suspended in a liquid polymer dope. Although the polymer solvents caused some damage to the MOF-74 crystals, the remaining crystals were still used for a subsequent growth procedure, resulting in MOF-74+Torlon monoliths. By applying a secondary growth technique, they created a MOF film-coated monolith, which allowed the material to regain its porosity and surface area. This process led to a five-fold increase in the isothermal adsorption capacity (2.5 mmol/g) at a temperature and pressure of 25°C and 1 bar CO_2, respectively, compared to the original MOF-74@Torlon monolith. The study highlighted the potential of 3D-printed MOF-based adsorbents, showing a noticeable increment in diffusivity, even though the ability for the CO_2 adsorption was lower than that of the parent MOF powders. This work opens up exciting opportunities for further research into secondary growth methods across different types of MOFs. The aim of employing the solvothermal synthesis method was to explore the formation of UTSA-16 MOF using a cobalt metal precursor and DIW 3D-printed kaolin monolith structures, as presented in **Figure 2.7 (b)** [47]. By adjusting the concentration of the metal precursor, it has become possible to acquire MOF loadings of up to 90% by weight. To improve the mechanical strength and exceed the performance of the self-standing UTSA-16-based monolith, cobalt in its oxide form was incorporated into the kaolin support. Since metal seeds were embedded within the UTSA-16 crystals rather than being placed on the surface, the surface area as well as the pore volume of the material were similar to those of the UTSA-16 powder. It has been observed that the kaolin-based monolith demonstrated comparable performance to the UTSA-16 powder, with a CO_2 adsorption capacity of the order of 3.1 mmol/g at a temperature and pressure of 25°C and 1 bar, respectively. The success of this approach was further validated through dynamic adsorption tests conducted under different binary gas mixtures and flow conditions. These tests highlighted the ability to create customised adsorbent structures with efficient mechanical properties and adsorption capacity.

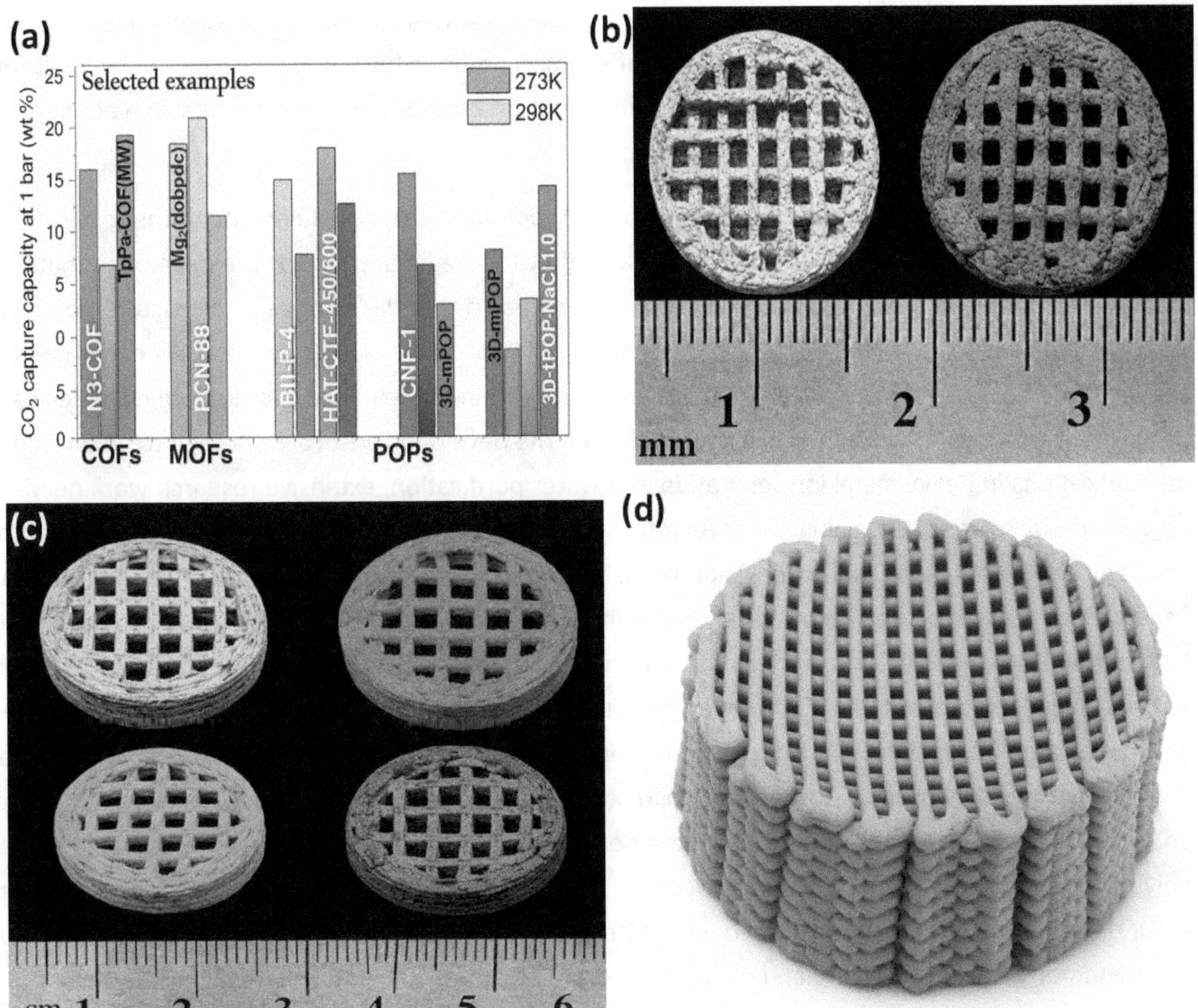

Figure 2.7: (a) Bar chart for the CO_2 capturing through different materials at two different temperatures, (b) Visualisation of before and after Co-kaolin monolith during the UTSA-16 formation, (c) Optical representation of the CD-printed HKUST-1@ Torlon and MOF-74@Torlon, (d) 3D-printed UTSA monolith with a diameter of 2.5 cm. Copyright Credit: Ref. [47], Applied Materials & Interfaces, ACS, 2018, Ref. [49], Industrial & Engineering Chemistry Research, ACS, 2020, Ref. [48]. Chemical Engineering Journal, Elsevier, 2020)

The ability of CO_2 to get adsorbed in the moisturised atmosphere is a crucial factor for separating it from flue gas. For resolving the issue, the adsorption capability of several gases, such as N_2, CH_4, CO_2, and water vapour, was studied by employing DIW 3D-printed UTSA-16 monoliths made with a non-aqueous ink containing boehmite and hydroxypropyl cellulose as binders, structure depicted in **Figure 2.7 (d)** [48]. While analysing the activity, it has been observed that the adsorption capacities of the CO_2 and H_2O monolith were about 60% of those observed in the UTSA-16 powder, thereby the tests revealed the monolith's strong ability to capture CO_2 selectively. Even at over 100% relative humidity, the material maintained stable CO_2 adsorption, demonstrating its potential for capturing CO_2 from humid air and showing promise for DAC applications. The process of advancements keeps on going, and different studies pertaining to different conditions have been carried out [49]. Different researchers have carried out different studies and tried to acquire the maximum possible efficient for the adsorption of the CO_2.

A composite adsorbent, made from a mixture of sodium alginate, cellulose nanofibrils, and COFs, was also 3D printed using DIW to create cellulose and COF-based material. For the developed material, the CO_2 adsorption capability was carried out, and it revealed that the synthesised structure possesses an adsorption capacity of 19.9 mg/g. This material is noteworthy because it successfully minimises the agglomeration of COF nanoparticles by embedding them within the cellulose matrix, forming molecularly linked interfaces and adsorption-active sites. Although the developed adsorbent proved to be promising for applications in heavy metal ion separation and water purification, extensive research work needs to be carried out related to scalability and long-term stability. A proper understanding of the adsorption mechanics and kinetics could provide valuable insights into its potential for gas separation applications. Recent studies have focused on adsorbent monoliths made by combining cellulose acetate with Zeolite 13X and MOF MIL-101(Cr) for efficient CO_2 capture from air at sub-ambient temperatures [50]. Zeolite 13X and microcrystals MIL-101(Cr) were successfully incorporated into the monoliths, which featured interpenetrated macroporous polymeric structures. This demonstrated the versatility of solution-based additive synthesis in creating sorbent monoliths with various particle sizes and densities. To further enhance CO_2 adsorption, the composite of the developed MIL-101 (Cr) was treated with the branched poly(ethylenimine) (PEI). It has been observed that the adsorption studies showed promising results, with a CO_2 uptake of 0.95 mmol/g for the monolith after 14 temperature swing cycles, and a CO_2 absorption of 1.05 mmol/g at a temperature and concentration of −20°C and 400 ppm CO_2, respectively, after being regenerated at 60°C.

Breakthrough studies conducted under both dry and humid (70% RH) conditions showed that the CO_2 absorption capacity of the monolith was 0.60 mmol/g at 25°C. This showed an increment in the absorption to 1.05 mmol/g at −20°C under dry conditions, and 1.43 mmol/g at −20°C under humid conditions. However, challenges remain, such as the relatively low MIL-101(Cr) loading capability (<70 wt%) and the delicacy of the monolith structures. These factors require further investigation into more complex geometries and scaling up the process. While the adsorption kinetics of the monolith are comparable to those of the powder, practical use of DAC sorbents at lower temperatures requires optimisation of desorption conditions and energy consumption. Achieving this will necessitate multidisciplinary research, particularly in the engineering of materials and the mechanical aspects of 3D printing [14].

CO_2 Capturing Using Membranes: Gas separation membranes are widely used for isolating gases, particularly in applications like flue gas treatment [51]. These membranes work by permitting one component of the gas mixture to pass through more easily while retaining the other, attributed to their semipermeable barrier and nanoporous structure. Two key factors that define the performance of a membrane are permeance (or permeability) and selectivity. Selectivity refers to how pure the gas component that passes through the membrane is, while permeability is defined as the rate at which the gas moves through the membrane, which in turn affects the pressure drop across it. Permeability refers to the flow rate of gas through the membrane, which relies on the pressure of the transmembrane and the area of the membrane. It's usually expressed in gas permeation units (GPU); the value of one GPU is defined as 3.35×10^{-1} $mols^{-1}$ m^{-2} Pa^{-1} [52]. Ideally, both permeability and selectivity should be high in a membrane, but in general, there's often a trade-off among them. It has been observed that membranes with high permeability generally have lower selectivity, and those with high selectivity tend to have lower permeability.

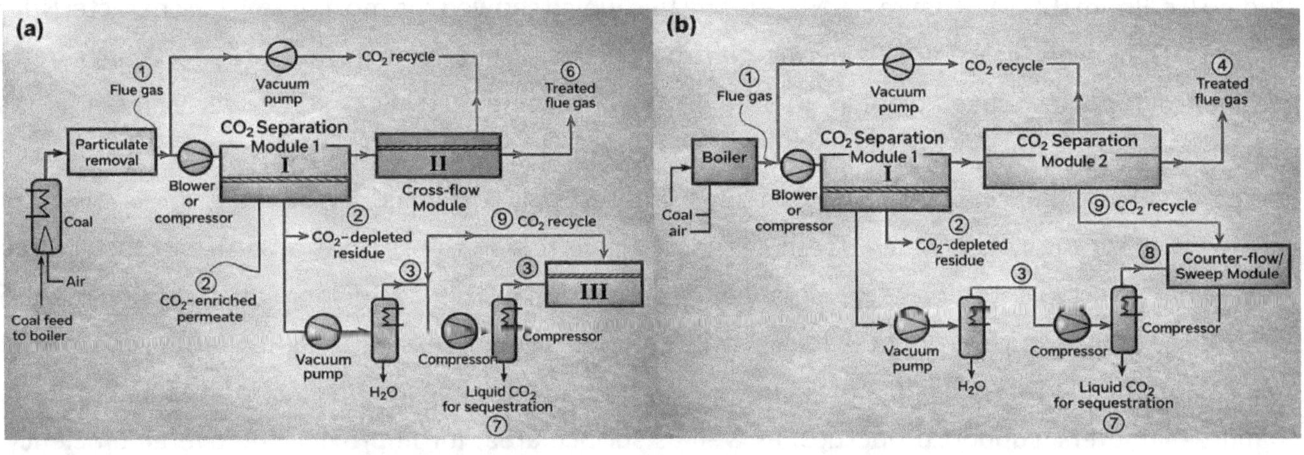

Figure 2.8: (a) Flow chart for a two-step vacuum membrane method used to extract and store CO_2 from coal-fired power station flue gas. (b) A two-step counter-flow/sweep membrane procedure intended to extract and store CO_2 from coal-fired power stations' flue gas is depicted in this simplified flow diagram. Copyright Credit: (Ref [53], Journal of Membrane Science, Elsevier, 2010)

One of the main advantages of membranes is their capability to provide continuous processing without any requirement for frequent rejuvenation. However, most of today's advanced membranes are made from polymers, which offer high selectivity but relatively low permeability. As a result, they require significant pressurisation, typically ranging from 7 to 13 bars. The schematic representation of the membrane processing units is shown in **Figure 2.8** [53]. Additionally, since flue gas contains a naturally low concentration of CO_2, achieving CO_2 purity comparable to amine scrubbing often requires multiple processing steps. This increases operational costs and limits the broader applicability of these membranes [54]. As a result, consequential efforts have been made toward fabricating membranes that offer both high permeability and selectivity. Notable achievements in this area include the creation of two-dimensional (2D) and three-dimensional (3D) membranes using porous graphene, other materials, and techniques, which provide permeability values a million times as compared to those of traditional polymer membranes.

Additionally, mixed-matrix membranes (MMMs) have shown promising results. Given these achievements, the investigation of MMMs. and graphene-based membranes proves to possess great potential for the future [55]. In this regard, some of the glimpses that have been made in generating the 3D membranes for CO_2 separation will also be discussed.

There are not many applications of 3D printing in CO_2 separation using ionic liquids (ILs) and membranes. In earlier studies, digital light processing (DLP) was used to print polymeric structures by copolymerising and trapping six commercial ILs [56]. A photorheological assessment, which evaluated the impact of ILs on monomer polymerisation kinetics, found that three of the selected ILs did not affect the polymerisation process. This highlighted the importance of optimising the exposure time of each layer during the printing process. With an IL content of 0.4 mmol/g, the developed materials were tested as assisted ionic liquid membranes to be employed for CO_2 capture. Among the different studied ILs, Bmim [BF_4] and Bmim [Tf_2N] were the two which showed the most promising characteristics owing to their high diffusion and permeation properties. While high surface-area-to-volume ratio gyroid structures have been created, their ability to perform effectively is still under evaluation. It's also important to conduct a broader analysis of the advantages and limitations of the current approach compared to conventional absorption systems. Interestingly, the majority of the printed structure consists of photo-polymer, which doesn't play an active role in the separation process, resulting in a relatively low concentration of ILs. To enhance separation efficiency, further research is needed to explore the structures developed with 3D printed materials possessing a higher proportion of the active component. This highlights the requirement for a comprehensive analysis of IL concentration, channel diameters, support properties, as well as surface area, for improving the overall efficiency. For explicit exploration of the potential pertaining to 3D printing for membrane development, the DLP technique was employed to print light-sensitive membranes using azobenzene dyes and PEGDA as precursors for the base polymer matrix [57]. Dyes methyl red and dispersed red 1 methacrylate were incorporated to allow for tunable permeability. The laser-induced azo photoisomerisation increased the membrane's free volume by 70%, boosting CO_2 permeability. However, the existence of water molecules in humid environments reduced CO_2 permeation. This underscores the need for further research, especially in flue gas settings, to account for other pollutants and the presence of moisture in the gas stream.

An alternative approach focused on MOFs utilised electrospray technology to create thin film composite membranes with a mixed matrix[58]. A mixture of HKUST-1 (a copper-based MOF) and Polymers of Intrinsic Microporosity (PIM-1) was prepared at different concentrations. For this, the precursor, which is 300 mg of HKUST-1 with 1.7 g of PIM-1 were mixed with 98 g of chloroform, then diluted to acquire the required concentrations of MOF. The concentration varied from 0.1% to 0.5% by weight. The acquired mixture was then electrosprayed on a pan 400 support, with the number of lamination scans adjusted to control the thickness of the selective layer. As the membrane thickness increases, CO_2/N_2 selectivity improves, particularly at lower CO_2 concentrations.

In comparison, analysing for the reason, it has been observed that this phenomenon is observed due to the presence of minor defects, which is a key difference between the developed membranes and the already present thinner membranes. In contrast to the pure PIM-1 membrane, the 0.5% by weight HKUST-1/PIM-1 TFC-MMM exhibited a significant four-fold increase in CO_2 permeance. Furthermore, to achieve better mechanical strength, selectivity, and permeance, adjustments to the composition and printing parameters are necessary.

While 3D printing has received significant attention for creating membranes used in desalination and water purification, its potential for CO_2 separation membranes has largely been unexplored[59]. However, in recent years, significant progress has been made in developing 3D-printed membranes for CO_2 capture. Early research paved the way for functionalised membranes by exploring the use of 3D printing to precisely control morphology and structure, as well as the ability to print ionic liquids for CO_2 collection. More recent studies have focused on composite membranes based on mixed-matrix thin film made from MOFs and light-responsive materials. Research is currently underway on mixed-matrix membranes and polymer composites created using 3D printing. The main aim of the studies will be to improve membrane efficiency, scalability, and affordability to ensure widespread industrial adoption. Recent advancements have made it possible to develop multifunctional membranes capable of selectively separating gases, purifying water, and exchanging ions. For further enhancing the membrane performance along with the properties, several efforts are being made in analysing new materials, such as infused nanoparticles and responsive polymers. While there have been studies on 3D-printed microchannels for enhancing the turbulence in membrane-based CO_2 absorption, more research is needed on the printed membrane materials themselves[60]. However, the studies presented here show a promising direction for using 3D printing to revolutionise selective CO_2 separation. By incorporating ILs and beneficial fillers into the printed membranes, this approach holds great potential.

CO_2 Conversion Through Catalysts: The role of 3D printing in catalysis is becoming increasingly important, especially with the rapid advancement of the technology and its potential to reduce energy consumption and operating costs. This shift is driven by 3D printing's ability to precisely control the shape and optimise the distribution of catalytic materials. Given the significant economic challenges and drawbacks associated with storing captured CO_2, the real challenge in addressing the CO_2 issue is not just capturing or separating it, but also converting it into useful and practical forms, thus preventing its unintended accumulation. The conversion of captured CO_2 through catalysis, either alongside capture or through a sequence of capture and usage processes, is a vital solution to this challenge. Particularly, if the by-products from the conversion process possess some commercial value, it offers an opportunity to recover the costs of both capture and conversion, achieving the dual goals of economic feasibility and environmental impact reduction. Recently, 3D printing has gained attention among researchers for creating catalysts and dual-functional materials (DFMs) for CO_2 mitigation. These combine CO_2 capture followed by its ensuing conversion within a developed material, thereby effectively serving two functions [14].

In this direction, CO_2 methanation is considered to be the most promising solution. It is an exothermic process with a high gas space velocity. To maintain their catalytic effectiveness, catalysts with higher heat and mass transfer properties need to be developed. In a groundbreaking study on 3D printing catalysts for CO_2 methanation, stainless steel substrates were printed using DIW and then coated with a Ni/Al_2O_3 layer through wash coating. The coating of Ni/Al_2O_3 is employed for making the surface catalytically active [61]. The catalytic monoliths, especially those with zigzag channels, showed improved efficiency in the methanation process after being activated at 450°C under controlled H_2/He gas flow. The enhancement in the efficiency was due to the zigzag channels, which allow long-duration contacts among the gas reactants and the surface of the catalyst. Compared to powdered catalysts, these monoliths resulted in significantly higher conversion rates. Throughout the methanation process performed at 350°C, the structured catalysts effectively managed the temperature increase and maintained stability over a challenging 53-hour evaluation. This marked the first successful usage of 3D printing technology for CO_2 conversion, accentuating the requirement for further exploration to fully optimise these catalytic materials and thoroughly assess their stability across various operating conditions. In another research work, copper supports were analysed, which were printed by employing the DIW technique followed by the coating with a Ni/Al_2O_3 catalyst, with cylindrical structure as shown in prototype **Figure 2.8 (a)** [62]. The effects of sintering temperature and environment were explored using both traditional and pulsed electrical current sintering (PEC) techniques. It has been observed that the PEC sintering process made oxidation more likely; this could be reduced by modifying the sintering temperature. The CO_2 methanation at temperatures between 280°C and 500°C showed that the structured catalysts performed better than their powdered counterparts in terms of both CO_2 conversion and CH_4 productivity. Notably, the structured catalysts showed resilience to contaminants like 10 ppm H_2S. While performance showed a significant improvement, full validation will require moving to pilot-scale testing employing the concentrated gas reactants under high-pressure setups. For the same, the efficiency of CO_2 methanation was investigated through a mini-pilot reactor, which showed a significant improvement in CH_4 output. Specifically, the 3D-printed sample produced 126.7 mmol/g.h, compared to 66.8 mmol/g.h from the powder sample. These enhanced results were attributed to several factors, including the reactor's geometry, large gas hourly space velocity, uniform heat dispersion, process layout, and effective pressure regulation [63]. Additionally, the material demonstrated stability throughout the process, with no hotspots observed and minimal carbon deposition. However, further research is needed to explore the relationship between reactor geometry, supports developed through 3D-printed, and their role in heat transfer and pressure drop, which are crucial for process intensification. In this regard, Middelkoop *et al.* [64] have studied the efficiency of CO_2 methanation. They have examined the process at various temperatures after Ni-Al_2O_3 catalysts were fabricated using commercial powder with a DIW 3D printer. The nozzle sizes used were 600 and 800 µm, aiming to evaluate the growth of the catalyst during CO_2 methanation. A pictorial representation with a nozzle is presented in **Figure 2.8 (b)**. The potential of this method was demonstrated by comparing the performance of the 3D-printed commercial catalysts (78% efficiency) with traditional pellet forms (58% efficiency), revealing significantly better CH_4 production efficiency for the 3D-printed catalysts. The distribution of Ni metals throughout the reaction was analysed using Operando XRD-Computerised Tomography, which provided detailed information on the catalyst's progression. However, for exploring the possible ways for the enhancement of the performance of printed catalysts, further investigation is needed on factors such as Ni loading, heat and mass transfer, followed by the comparative analysis pertaining to the effect of different catalytic supports. For optimisation of the heat transfer, the selective laser sintering technique was employed to 3D-print metallic honeycomb monoliths coated with Ni/CeO_2-ZrO_2 [65]. The metal honeycomb structures showed minimal deactivation over extended reaction times and achieved nearly cent percent CO_2 conversion at 900°C during the methane dry reforming (DRM) process. Compared to their conventional cordierite-supported counterparts, the metallic supports improved heat transport, thereby resolving the requirement of activation time in the process of DRM. For increasing the efficiency of the heat transfer, pseudo-gyroid-based stainless steel Fluid Guiding Elements (FGE) were 3D printed through the use of selective laser melting technology. The prototype is shown in **Figure 2.8 (c)** [66]. These structures were then anchored onto $MgAl_2O_4$, followed by an active catalyst coating of 15% Ni and 0.5% Ru before being employed for CO_2 methanation. It has been observed that the developed FGEs exhibit efficient catalytic activity, achieving 75% CO_2 conversion at a temperature of 400°C, due to a reduced tendency for centralised sintering and enhanced radial heat transport. Computational simulations were used to compute the optimal perquisites for the CO_2 methanation process. As the study continues, proper investigation into geometric factors as well as the substrate materials will be required before proceeding with the lab-scale testing process.

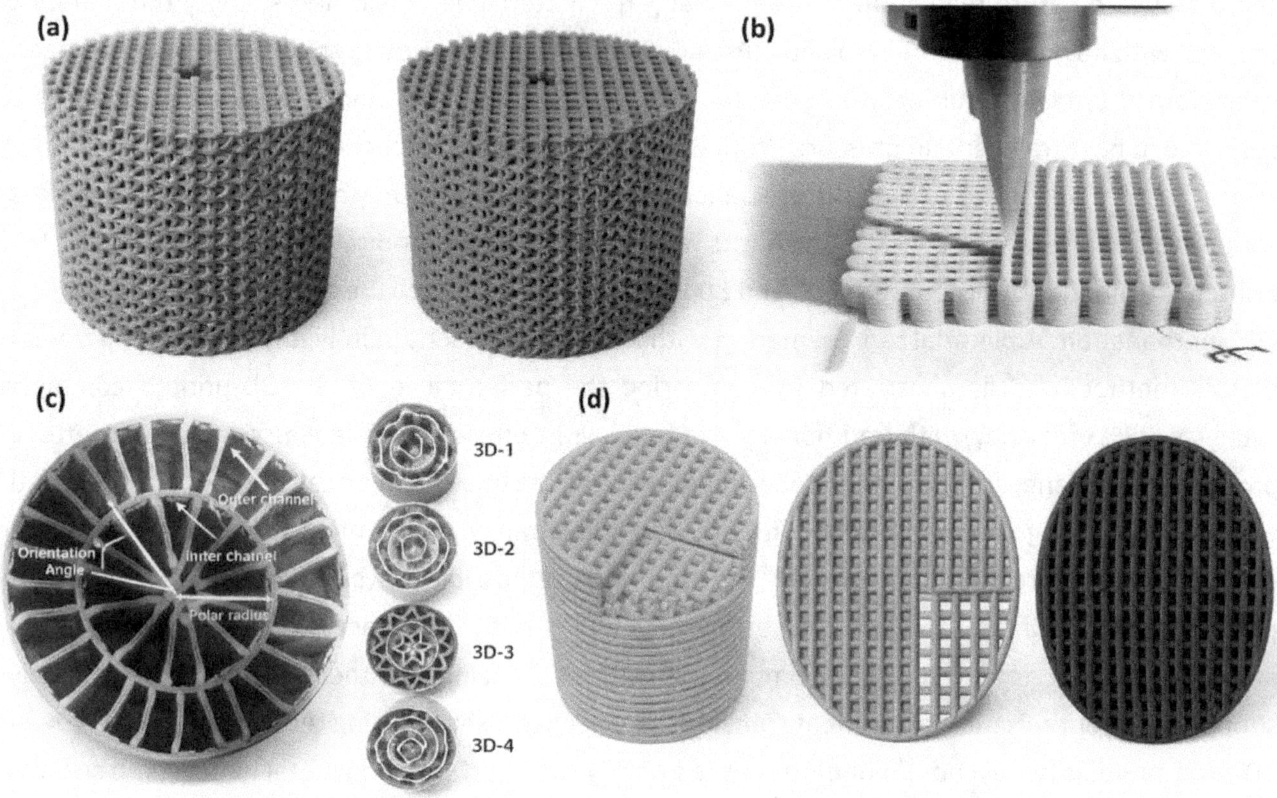

Figure 2.9: Different images of the 3D printed monoliths: (a) stainless steel and copper, (b) prototype catalyst for Ni/Al_2O_3, (c) Fluid guide element made of stainless steel employed for the CO_2 methanation, (d) Prototype models along with stack employed for Na-Fe@C. Copyright Credit: (Ref. [62], Chemical Engineering & Processing, Elsevier, 2018, Ref. [64], Journal of CO2 Utilization, Elsevier, 2019, Ref. [66], Energy Conversion and Management, Elsevier, 2022, Ref. [67], Applied Catalysis B, Elsevier, 2024)

Although it has been well established that placing catalysts on metal supports enhances heat transfer, there has been no comprehensive simulation presented in existing studies that demonstrates the progress of the temperature profile and associated heat transfer trajectory during the reaction process. While many researchers have investigated this at the material level, these changes become increasingly important when scaling up to pilot-sized reactors. Additionally, for a proper understanding of the mechanism of heat transfer, it is important to perform an experimental, detailed analysis of the phenomenon, which will be helpful in guiding efforts to improve the catalyst's thermal properties. Recently, Fe-based MOF monolithic catalysts were 3D-printed using DIW to understand better heat and mass transfer in CO_2 hydrogenation, shown in the prototype representation as **Figure 2.8 (d)** [67]. For the production of light olefins, the developed spiral-type design demonstrated excellent selectivity, achieving 52.6%. Besides that, it has been analysed through the mass transfer simulations that the spiral channels optimise the use of active sites and control carbon-chain growth by enhancing mass transfer, thereby minimising the accumulation of mediators and by-products at the catalytic interfaces. The catalytic interfaces with high electron density and minimal intermediate coverage promote light olefin synthesis by slowing down the coupling of carbon–carbon bonds and aiding in the desorption of olefins, which was confirmed through DFT simulations. Furthermore, the 3D geometry results in an enhancement of the heat transport, extending the catalyst's lifespan by reducing carbon deposition and inhibiting the agglomeration of active sites.

This research study leverages 3D printing to strategically improve the mass as well as heat transfer properties of monolithic-based catalysts. This phenomenon of enhancement has been observed in the spiral-channel CO_2 hydrogenation Fe-based catalysts that deliver high light olefin yields and extended catalytic lifetimes. Besides that, the simulation studies provide strong evidence that optimising mass transfer plays a crucial role in enhancing olefin processing by reducing over-hydrogenation and limiting the growth of carbon chains. At the same time, improved heat transfer confirms the long-standing reliability of the catalyst. Tackling the problem through this approach not only enhances the distribution of CO_2 hydrogenation products but also offers a versatile method for customising catalysts, which can be applied across a wide range of catalytic systems. For analysing the properties of the metal oxide catalysts, carbon monoliths with large surface area were 3D-printed through DIW technique and employed as reinforcement for the catalyst's active phase, that is NiO-CeO_2 (12 wt% Ni) [68]. Using the monolith fabrication, two structures were developed: a traditional honeycomb structure and a more advanced network of non-linear channels, which increases the diffusion of gas by producing disturbances along the path of gas reactants. It has been observed that the circular shape monoliths showed enhanced gas diffusion corresponding to the nanoparticles, which tends to improve the low-temperature (<300°C) CO_2 methanation process. However, at elevated temperatures (>300°C), where mass transfer limitations govern the reaction kinetics, the conventional catalysts resulted in more CH_4 creation. Furthermore, the microporosity exhibited by the monoliths allows the deeper penetration of the nanoparticles, easier agglomeration of the particles on the channel surfaces, along with the frequent accessibility to the reactant molecules. This contributed to excellent stability and recyclability. However, it is necessary to further investigate the longevity, durability and other important aspects of these catalysts by operating for a longer duration of time under different pressure conditions. In a separate study, the reverse water-gas shift (RWGS) reaction was used to convert CO_2 with molybdenum carbide-based catalysts, supported by Al_2O_3, by using the 3D-printing DIW technology. The results indicated that the catalyst's performance is significantly influenced by crystallite size and phase composition, both of which are affected by Mo loading [69]. Operando DRIFTS test revealed how hydroxyl species on Al_2O_3 play a role in the formation of bicarbonate, which then leads to the production of formates, CO, and H_2O. The catalyst with approximately 10 wt% Mo loading exhibits enhanced catalytic activity, producing the most CO at a rate of 17.6 mol/kg/h at a temperature and pressure of 873 K and 0.1 MPa, respectively, along with a velocity of gas production of 4000/h. For scale-up, the mechanical stability of the 3D Mo_xC/Al_2O_3 structures appeared promising. However, it is important to note that while coke deposition was not specifically identified as a major issue, the potential for coke formation could still pose challenges.

Furthermore, the suggested method might be expanded to additional transition metal carbides for a range of catalytic applications by choosing suitable metal precursors and improving the accessibility of metal centres in the printed sections. Together, these experiments show that achieving optimal catalytic effectiveness, along with anticipated product yield, requires precise adjustment of the metal oxide composition. These catalysts must be carefully designed for 3D printing, as the potential for pore blockage due to the coalescence of powder components could alter their intended functionality.

Although different materials have been discussed above, used for the conversion of CO_2, along with their progress and current scenarios. The discussion provided here is concise but insightful, so that readers will get a proper outlook of the different aspects currently going on in the field. Besides the above-discussed materials, due to advanced technology, some new progress has been made in the field of adsorbents. The newly developed adsorbents, which also exhibit a significant contribution to the field, include the 3D-printed zeolites, binder-free absorbents, and bifunctional materials developed using 3D printed technology.

2.5 Challenges with Advanced Materials for CO_2 Capture

In this chapter, we have explored various materials and their advancements in the field of CO_2 capture. It is evident that, with the progress of technologies like additive manufacturing and 3D printing, each material has entered a new phase, showing much greater improvements compared to their older counterparts. While all the materials discussed, such as adsorbents, membranes, and catalysts, demonstrate excellent features, their 3D-printed versions stand out for their pressure drop conditions, enhanced heat and mass transfer, as well as the flexibility in designing geometries. Evaluating the effectiveness of complex geometries in heat and mass transfer dynamics during adsorption and desorption requires further research and optimisation. While 3D printing holds significant potential for reducing CO_2 conversion, challenges remain, especially in optimising material formulations for printing. Each material requires a unique combination of active ingredients, binders, and additives, as there is no one-size-fits-all solution. Additionally, the weak sorption capacity of binder-based monoliths has impelled further investigation into binder-less alternatives, which, in some cases, have demonstrated comparable or even much better performance as compared to their powdered counterparts.

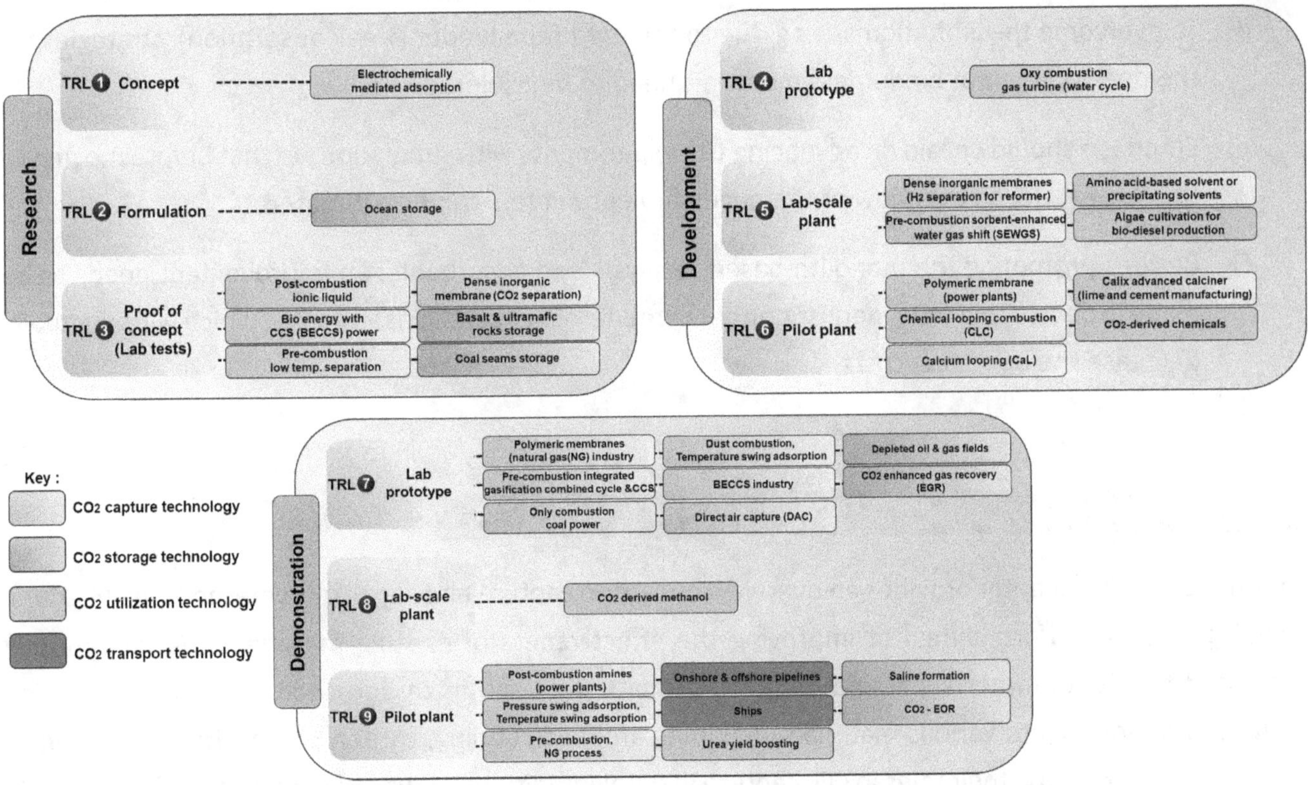

Figure 2.10: Technology Readiness Levels involved in the CCUS technology, depicting the progression from TRL (1–3), TRL (4–6), and TRL (7–9), respectively. Copyright Credit: (Ref. [13], Energy and Environmental Science, RSC, 2026)

Furthermore, the advancement of catalysts towards pilot-scale systems depends on the development of more complex structures and a thorough assessment of their catalytic efficiency. Scaling up to pilot-scale systems will result in further understanding of increasing the Technology Readiness Levels (TRL; **Figure 2.10**), and ensuring economic viability. Encouragingly, the cyclic stability observed in laboratory-scale experiments is auspicious. However, additional experiments and simulations are needed to confirm potential improvements in heat transfer with suitable material integration. When compared to traditional pelletised catalysts, monoliths offer significant benefits due to their reduced pressure drop and the synergistic advantages of improved heat and mass transfer, ultimately leading to lower operating costs. The concise summary for the future can be put as:

a) 3D printing, as well as additive manufacturing, needs to be implemented in an optimised way for carbon capturing. It will be possible to use these two technologies to develop respective useful geometries under optimised pressure, heat and mass transfer.

b) Through investigation, there needs to be done on exploration of the different geometric shapes with respect to the efficient adsorption–desorption dynamics.

c) For designing any material, we need different precursors such as active components, binder, and additives in the appropriate portions. As such, no universal formulation exists; emphasis needs to be made on developing advanced printing formulations.

d) To overcome the limitation arising due to the use of the binder (weak adsorption), strategies for the development of binder-less materials need to be explored.

e) Emphasis should be laid on advancing the experiments with simulation, so that before moving to the large scale, an overview of the necessary parameters can be estimated.

f) Besides promoting the monolith-based catalysts and adsorbents, an improvement needs to be made in developing the synergistic enhancements of heat and mass transfer, which will eventually decrease the operating costs.

2.6 Conclusion

In conclusion, this chapter provides an overview of carbon capture and separation methods, highlighting ongoing research efforts aimed at improving the effectiveness of existing materials. We have briefly discussed the need, importance, and progress of materials used in carbon capture technologies. The chapter also delves into various materials, including amine scrubbing, carbon composites, membranes, and catalysts, exploring their roles in CO_2 capture in detail. Additionally, we examined how advancements in additive manufacturing and 3D printing technologies have enhanced the effectiveness of these materials. The discussion highlights that most research in the field of 3D printing for CO_2 capture, separation, and conversion has taken place over the past many years, marking a significant step in the development of this technology. To fully understand and optimise capture and conversion processes, 3D-printed materials must be tested in a variety of industrially relevant environments. Computational modelling and simulations can aid in predicting and optimising the best experimental conditions and designs for monoliths. However, further experimental validation with different printed materials in real-world conditions is crucial, especially when simulating applications such as direct air capture in CO_2 removal or flue gas treatment. The primary focus is now on emerging technologies that could help to achieve net-zero emissions by 2050. In this regard, some startup companies have already advanced to demonstration facilities at tonne-scale and are set to expand further. With the expected rise in carbon prices due to strict regulations in the coming years, these innovative companies are likely to capture a significant share of the carbon capture market by providing more cost-effective solutions for both operation and installation.

REFERENCES

[1] Negro, C., Garcia-Ochoa, F., Tanguy, P., Ferreira, G., Thibault, J., Yamamoto, S., and Gani, R. (2018) WITHDRAWN: Barcelona Declaration – 10th World Congress of Chemical Engineering, 1–5 October 2017. *Chem. Eng. Res. Des.*, **129**, 34.

[2] Energy Information Administration (2011) International Energy Outlook 2011 - Energy Information Administration. *U.S. Energy Inf. Adm.*

[3] Vooradi, R., Bertran, M.O., Frauzem, R., Anne, S.B., and Gani, R. (2018) Sustainable chemical processing and energy-carbon dioxide management: Review of challenges and opportunities. *Chem. Eng. Res. Des.*, **131**, 440–464.

[4] Hare, B., Schaeffer, M., Rocha, M., and LDCs (2011) Science aspects of the 2°C and 1.5°C global goals in the Cancun Agreements. 1–21.

[5] Ashirov, T., and Coskun, A. (2024) Carbon Dioxide Capture: Current Status and Future Prospects. *Chimia (Aarau).*, **78** (6), 415–422.

[6] Peters, G.P., Marland, G., Le Quéré, C., Boden, T., Canadell, J.G., and Raupach, M.R. (2012) Rapid growth in CO_2 emissions after the 2008-2009 global financial crisis. *Nat. Clim. Chang.*, **2** (1), 2–4.

[7] Patel, H.A., Byun, J., and Yavuz, C.T. (2017) Carbon Dioxide Capture Adsorbents: Chemistry and Methods. *ChemSusChem*, **10** (7), 1303–1317.

[8] Liu, J., Thallapally, P.K., Mc Grail, B.P., Brown, D.R., and Liu, J. (2012) Progress in adsorption-based CO_2 capture by metal–organic frameworks. *Chem. Soc. Rev.*, **41** (6), 2308–2322.

[9] Pires, J.C.M., Martins, F.G., Alvim-Ferraz, M.C.M., and Simões, M. (2011) Recent developments on carbon capture and storage: An overview. *Chem. Eng. Res. Des.*, **89** (9), 1446–1460.

[10] Baker, R.W., and Low, B.T. (2014) Gas separation membrane materials: A perspective. *Macromolecules*, **47** (20), 6999–7013.

[11] Mc Keown, N.B., and Budd, P.M. (2006) Polymers of intrinsic microporosity (PIMs): Organic materials for membrane separations, heterogeneous catalysis and hydrogen storage. *Chem. Soc. Rev.*, **35** (8), 675–683.

[12] Rezaei, F., and Webley, P. (2009) Optimum structured adsorbents for gas separation processes. *Chem. Eng. Sci.*, **64** (24), 5182–5191.

[13] Madhusudhana M. Devadiga et al., Describing critical barriers in commercial deployment of CO2 capture and utilization technologies, Energy Environ. Sci., 2026,19, 743-790https://doi.org/10.1039/D6EE00134C

[14] Jivrakh, K.B., Kuppireddy, S., Dumée, L.F., Polychronopoulou, K., Abu Al-Rub, R.K., Alamoodi, N., and Karanikolos, G.N. (2024) A critical review on 3D-printed adsorbents, membranes, and catalysts for carbon dioxide capture, separation, and conversion. *J. Clean. Prod.*, **472**, 143522.

[15] Ali, M., Jha, N.K., Pal, N., Keshavarz, A., Hoteit, H., and Sarmadivaleh, M. (2022) Recent advances in carbon dioxide geological storage, experimental procedures, influencing parameters, and future outlook. *Earth-Science Rev.*, **225**, 103895.

[16] Pan, H., Yu, C., Suo, X., Yang, L., Cui, X., and Xing, H. (2023) Emerging porous materials for carbon dioxide adsorptive capture: progress and challenges. *Mater. Chem. Front.*, **7** (24), 6463–6482.

[17] Ben-Mansour, R., Habib, M.A., Bamidele, O.E., Basha, M., Qasem, N.A.A., Peedikakkal, A., Laoui, T., and Ali, M. (2016) Carbon capture by physical adsorption: Materials, experimental investigations and numerical modeling and simulations - A review. *Appl. Energy*, **161**, 225–255.

[18] Rochelle, G.T. (2009) Amine Scrubbing for CO_2 Capture. *Science (80).*, **325** (5948), 1652–1654.

[19] Jagdeep Singh et al., Tailoring carbon nanomaterial architectures for CO2 capture: structure–property relationships, surface engineering, and future perspectives, Materials Advances, 7, 6,2026,3031-3071, https://doi.org/10.1039/d5ma01408e

[20] Mutch, G.A., Shulda, S., McCue, A.J., Menart, M.J., Ciobanu, C. V., Ngo, C., Anderson, J.A., Richards, R.M., and Vega-Maza, D. (2018) Carbon Capture by Metal Oxides: Unleashing the Potential of the (111) Facet. *J. Am. Chem. Soc.*, **140** (13), 4736–4742.

[21] Krödel, M., Landuyt, A., Abdala, P.M., and Müller, C.R. (2020) Mechanistic Understanding of CaO-Based Sorbents for High-Temperature CO_2 Capture: Advanced Characterization and Prospects. *ChemSusChem*, **13** (23), 6259–6272.

[22] Kierzkowska, A.M., Pacciani, R., and Müller, C.R. (2013) CaO-based CO_2 sorbents: From fundamentals to the development of new, highly effective materials. *ChemSusChem*, **6** (7), 1130–1148.

[23] Varghese, A.M., and Karanikolos, G.N. (2020) CO_F capture adsorbents functionalized by amine – bearing polymers: A review. *Int. J. Greenh. Gas Control*, **96**, 103005.

[24] Wick-Joliat, R., Weisshar, F.B., Gorbar, M., Meier, D.M., and Penner, D. (2024) CO_2 Capture with Polyethylenimine Supported on 3D-Printed Porous SiO_2 Structures. *Mater. 2024, Vol. 17, Page 2913*, **17** (12), 2913.

[25] Thakkar, H., Eastman, S., Al-Mamoori, A., Hajari, A., Rownaghi, A.A., and Rezaei, F. (2017) Formulation of Aminosilica Adsorbents into 3D-Printed Monoliths and Evaluation of Their CO_2 Capture Performance. *ACS Appl. Mater. Interfaces*, **9** (8), 7489–7498.

[26] Sluijter, S.N., Boon, J., James, J., Krishnamurthy, S., Lind, A., Blom, R., Andreassen, K.A., Cormos, A.M., Sandu, V.C., and de Boer, R. (2021) 3D-printing of adsorbents for increased productivity in carbon capture applications (3D-CAPS). *Int. J. Greenh. Gas Control*, **112**, 103512.

[27] Jacobs, M., De Vos, Y., and Rombouts, M. (2024) 3D printed electrically conductive amine based sorbent for CO_2 capture. *Int. J. Greenh. Gas Control*, **132**, 104064.

[28] Sircar, S., Golden, T.C., and Rao, M.B. (1996) Activated carbon for gas separation and storage. *Carbon N. Y.*, **34** (1), 1–12.

[29] Song, K.S., Fritz, P.W., and Coskun, A. (2022) Porous organic polymers for CO_2 capture, separation and conversion. *Chem. Soc. Rev.*, **51** (23), 9831–9852.

[30] Wang, J., Pu, Q., Ning, P., and Lu, S. (2021) Activated carbon-based composites for capturing CO_2: a review. *Greenh. Gases Sci. Technol.*, **11** (2), 377–393.

[31] Nguyen, T.S., Dogan, N.A., Lim, H., and Yavuz, C.T. (2023) Amine Chemistry of Porous CO_2 Adsorbents. *Acc. Chem. Res.*, **56** (19), 2642–2652.

[32] Yang, X., Zhao, L., Yang, S., and Xiao, Y. (2013) Investigation of natural CaO-MgO sorbent for CO_2 capture. *Asia-Pacific J. Chem. Eng.*, **8** (6), 906–915.

[33] Zafanelli, L.F.A.S., Henrique, A., Steldinger, H., Diaz de Tuesta, J.L., Gläsel, J., Rodrigues, A.E., Gomes, H.T., Etzold, B.J.M., and Silva, J.A.C. (2022) 3D-printed activated carbon for post-combustion CO_2 capture. *Microporous Mesoporous Mater.*, **335**, 111818.

[34] Comroe, M.L., Kolasinski, K.W., and Saha, D. (2022) Direct Ink 3D Printing of Porous Carbon Monoliths for Gas Separations. *Molecules*, **27** (17), 5653.

[35] Verougstraete, B., Schoukens, M., Sutens, B., Vanden Haute, N., De Vos, Y., Rombouts, M., and Denayer, J.F.M. (2022) Electrical swing adsorption on 3D-printed activated carbon monoliths for CO_2 capture from biogas. *Sep. Purif. Technol.*, **299**, 121660.

[36] McHale, C.M., Stegemoller, C.R., Hashim, M.I., Wang, X., and Miljanić, O. (2019) Porosity and Guest Inclusion in Cyclobenzoin Esters. *Cryst. Growth Des.*, **19** (2), 562–567.

[37] Justin, A., Espín, J., Pougin, M.J., Stoian, D., Schertenleib, T., Mensi, M., Kochetygov, I., Ortega-Guerrero, A., and Queen, W.L. (2024) Post-Synthetic Covalent Grafting of Amines to NH2-MOF for Post-Combustion Carbon Capture. *Adv. Funct. Mater.*, **34** (7), 2307430.

[38] Bai, L., Gao, Q., and Zhao, Y. (2016) Two fully conjugated covalent organic frameworks as anode materials for lithium ion batteries. *J. Mater. Chem. A*, **4** (37), 14106–14110.

[39] Wei, H., Chai, S., Hu, N., Yang, Z., Wei, L., and Wang, L. (2015) The microwave-assisted solvothermal synthesis of a crystalline two-dimensional covalent organic framework with high CO_2 capacity. *Chem. Commun.*, **51** (61), 12178–12181.

[40] Li, H., Dilipkumar, A., Abubakar, S., and Zhao, D. (2023) Covalent organic frameworks for CO_2 capture: from laboratory curiosity to industry implementation. *Chem. Soc. Rev.*, **52** (18), 6294–6329.

[41] Modak, A., and Jana, S. (2019) Advancement in porous adsorbents for post-combustion CO_2 capture. *Microporous Mesoporous Mater.*, **276**, 107–132.

[42] Ding, M., Flaig, R.W., Jiang, H.L., and Yaghi, O.M. (2019) Carbon capture and conversion using metal-organic frameworks and MOF-based materials. *Chem. Soc. Rev.*, **48** (10), 2783–2828.

[43] McDonald, T.M., Lee, W.R., Mason, J.A., Wiers, B.M., Hong, C.S., and Long, J.R. (2012) Capture of carbon dioxide from air and flue gas in the alkylamine-appended metal-organic framework mmen-Mg 2(dobpdc). *J. Am. Chem. Soc.*, **134** (16), 7056–7065.

[44] Li, J.R., Yu, J., Lu, W., Sun, L.B., Sculley, J., Balbuena, P.B., and Zhou, H.C. (2013) Porous materials with pre-designed single-molecule traps for CO_2 selective adsorption. *Nat. Commun.*, **4** (1), 1–8.

[45] Abuzeid, H.R., EL-Mahdy, A.F.M., and Kuo, S.W. (2021) Covalent organic frameworks: Design principles, synthetic strategies, and diverse applications. *Giant*, **6**, 100054.

[46] Zhuo, Y., Hariyani, S., Armijo, E., Abolade Lawson, Z., and Brgoch, J. (2020) Evaluating Thermal Quenching Temperature in Eu^{3+}-Substituted Oxide Phosphors via Machine Learning. *ACS Appl. Mater. Interfaces*, **12** (5), 5244–5250.

[47] Lawson, S., Al-Naddaf, Q., Krishnamurthy, A., Amour, M.S., Griffin, C., Rownaghi, A.A., Knox, J.C., and Rezaei, F. (2018) UTSA-16 Growth within 3D-Printed Co-Kaolin Monoliths with High Selectivity for CO_2/CH_4, CO_2/N_2, and CO_2/H_2 Separation. *ACS Appl. Mater. Interfaces*, **10** (22), 19076–19086.

[48] Grande, C.A., Blom, R., Middelkoop, V., Matras, D., Vamvakeros, A., Jacques, S.D.M., Beale, A.M., Di Michiel, M., Anne Andreassen, K., and Bouzga, A.M. (2020) Multiscale investigation of adsorption properties of novel 3D printed UTSA-16 structures. *Chem. Eng. J.*, **402**, 126166.

[49] Lawson et al., Development of 3D-Printed Polymer-MOF Monoliths for CO2 Adsorption,Industrial & Engineering Chemistry Research, Ind. Eng. Chem. Res. 2020, 59, 15, 7151–7160, https://doi.org/10.1021/acs.iecr.9b05445.

[50] Wang, Y., Rim, G., Song, M.G., Holmes, H.E., Jones, C.W., and Lively, R.P. (2024) Cold Temperature Direct Air CO_2 Capture with Amine-Loaded Metal-Organic Framework Monoliths. *ACS Appl. Mater. Interfaces*, **16** (1), 1404–1415.

[51] Powell, C.E., and Qiao, G.G. (2006) Polymeric CO_2/N_2 gas separation membranes for the capture of carbon dioxide from power plant flue gases. *J. Memb. Sci.*, **279** (1–2), 1–49.

[52] Stern, S.A. (1968) The "barrer" permeability unit. *J. Polym. Sci. Part A-2 Polym. Phys.*, **6** (11), 1933–1934.

[53] Merkel, T.C., Lin, H., Wei, X., and Baker, R. (2010) Power plant post-combustion carbon dioxide capture: An opportunity for membranes. *J. Memb. Sci.*, **359** (1–2), 126–139.

[54] Roussanaly, S., Anantharaman, R., Lindqvist, K., Zhai, H., and Rubin, E. (2016) Membrane properties required for post-combustion CO2 capture at coal-fired power plants. *J. Memb. Sci.*, **511**, 250–264.

[55] Celebi, K., Buchheim, J., Wyss, R.M., Droudian, A., Gasser, P., Shorubalko, I., Kye, J. Il, Lee, C., and Park, H.G. (2014) Ultimate permeation across atomically thin porous graphene. *Science (80).*, **344** (6181), 289–292.

[56] Gillono, M., Chiappone, A., Mendola, L., Gomez, M.G., Scaltrito, L., Pirri, C.F., and Roppolo, I. (2019) Study on the printability through digital light processing technique of ionic liquids for CO2 capture. *Polymers (Basel).*, **11** (12), 1932.

[57] Gillono, M., Roppolo, I., Frascella, F., Scaltrito, L., Pirri, C.F., and Chiappone, A. (2020) CO_2 permeability control in 3D printed light responsive structures. *Appl. Mater. Today*, **18**, 100470.

[58] Elsaidi, S.K., Ostwal, M., Zhu, L., Sekizkardes, A., Mohamed, M.H., Gipple, M., McCutcheon, J.R., and Hopkinson, D. (2021) 3D printed MOF-based mixed matrix thin-film composite membranes. *RSC Adv.*, **11** (41), 25658–25663.

[59] Khalil, A., Ahmed, F.E., and Hilal, N. (2021) The emerging role of 3D printing in water desalination. *Sci. Total Environ.*, **790**, 148238.

[60] Ho, C.D., Chen, L., Tu, J.W., Lin, Y.C., Lim, J.W., and Chen, Z.Z. (2023) Investigation of CO_2 Absorption Rate in Gas/Liquid Membrane Contactors with Inserting 3D Printing Mini-Channel Turbulence Promoters. *Membranes (Basel)*, **13** (12), 899.

[61] Danaci, S., Protasova, L., Lefevere, J., Bedel, L., Guilet, R., and Marty, P. (2016) Efficient CO_2 methanation over Ni/Al_2O_3 coated structured catalysts. *Catal. Today*, **273**, 234–243.

[62] Danaci, S., Protasova, L., Snijkers, F., Bouwen, W., Bengaouer, A., and Marty, P. (2018) Innovative 3D-manufacture of structured copper supports post-coated with catalytic material for CO2 methanation. *Chem. Eng. Process. - Process Intensif.*, **127**, 168–177.

[63] Danaci, S., Protasova, L., Middelkoop, V., Ray, N., Jouve, M., Bengaouer, A., and Marty, P. (2019) Scaling up of 3D printed and Ni/Al_2O_3 coated reactors for CO_2 methanation. *React. Chem. Eng.*, **4** (7), 1318–1330.

[64] Middelkoop, V., Vamvakeros, A., De Wit, D., Jacques, S.D.M., Danaci, S., Jacquot, C., De Vos, Y., Matras, D., Price, S.W.T., and Beale, A.M. (2019) 3D printed Ni/Al_2O_3 based catalysts for CO2 methanation-a comparative and operando XRD-CT study. *J. CO_2 Util.*, **33**, 478–487.

[65] Agueniou, F., Vidal, H., de Dios López, J., Hernández-Garrido, J.C., Cauqui, M.A., Botana, F.J., Calvino, J.J., Galvita, V. V., and Gatica, J.M. (2021) 3D-printing of metallic honeycomb monoliths as a doorway to a new generation of catalytic devices: the Ni-based catalysts in methane dry reforming showcase. *Catal. Commun.*, **148**, 106181.

[66] González-Castaño, M., Baena-Moreno, F., Carlos Navarro de Miguel, J., Miah, K.U.M., Arroyo-Torralvo, F., Ossenbrink, R., Odriozola, J.A., Benzinger, W., Hensel, A., Wenka, A., and Arellano-García, H. (2022) 3D-printed structured catalysts for CO_2 methanation reaction: Advancing of gyroid-based geometries. *Energy Convers. Manag.*, **258**, 115464.

[67] Wang, Y., Lin, S., Li, M., Zhu, C., Yang, H., Dong, P., Lu, M., Wang, W., Cao, J., Liu, Q., Feng, X., Hu, H., Tsubaki, N., and Wu, M. (2024) Boosting CO2 hydrogenation of Fe-based monolithic catalysts via 3D printing technology-induced heat/mass-transfer enhancements. *Appl. Catal. B Environ.*, **340**, 123211.

[68] Martínez-López, I., Martínez-Fuentes, J.C., Bueno-Ferrer, J., Davó-Quiñonero, A., Guillén-Bas, E., Bailón-García, E., Lozano-Castelló, D., and Bueno-López, A. (2024) Structural design and particle size examination on NiO-CeO_2 catalysts supported on 3D-printed carbon monoliths for CO2 methanation. *J. CO2 Util.*, **81**, 102733.

[69] Pajares, A., Andrade-Arvizu, J., Jain, D., Monai, M., Lefevere, J., de la Piscina, P.R., Homs, N., and Michielsen, B. (2024) Exploring the 3D printing of molybdenum carbide-based catalysts for the reverse water gas shift reaction: A multi scale study. *Chem. Eng. J.*, **482**, 149048.

Chapter 3

HYBRID MATERIALS AND SURFACE ENGINEERING FOR CARBON CAPTURE

Abstract: The increasing demand for carbon capturing and storage has made a lot of progress in developing different materials. Through the fact that different techniques have been employed for designing the materials, cost-effectiveness, efficiency, and energy efficiency are some of the prominent factors that still act as barriers to achieving the goal. Among the different ways of capturing, adsorption is considered one of the easiest ways owing to its low cost, ease, and wide range of usage. With the aim of achieving efficiency in different aspects, different approaches are being applied. Among the different approaches that have been employed for modifications, surface engineering is the most viable approach that has been employed and has been proven to be effective. In this chapter, an overview of the impact of surface engineering and its different aspects will be discussed in detail.

3.1 Introduction

The concentration of CO_2 in the atmosphere has been continuously rising for many centuries and has currently marked an increase of about 40% since 1960 [1]. It has been assessed that in the atmosphere, CO_2 is present at an average concentration of 421 ppm [2]. For the environmentally friendly development, different world organisations have taken different steps, especially the United Nations has set regulations for decreasing the CO_2 emission level by 45% till 2030 as compared to the amount released in 2010. Besides that, the soul efforts are being directed to develop the technology so that by 2050, we can have a transition to the technology with net-zero emissions [3]. While assessing the different sources contributing to the CO_2 emission, they can be broadly classified into the following three different categories **(Figure 3.1)**: (i) point sources which mainly includes emission from coal-based power plants, different kilns, natural boiler with gas as fuel, and different chemical producing industries; (ii) mobile sources which mainly includes the different sources of transportation; and (iii) daily usage residential as well as commercial sources, which includes fuels based lightening, heating, and cooling. Among the different prerequisites of daily life, energy in different forms is one of the main pillars of the modern generation. Among the different forms of energy, power is of utmost important. However, among the different available sources, fossil fuel-based energy-producing industries fulfil about 70% of the power requirement demands [4]. In this regard, we need to account of the different emission sources that are posing thread to the living atmosphere while ensuring the fulfilment of needs and taking care of the future rising demands for the energy. Thus, some the steps towards this process will be (i) curtailing the use of fossil fuels while maintaining the energy outputs, (ii) focusing on enhancing the energy-efficiency output of the power plants, (iii) progress the system of carbon capturing and storage (CCS), and (iv) developing and progressing the technology for furnishing the nuclear and solar energy.

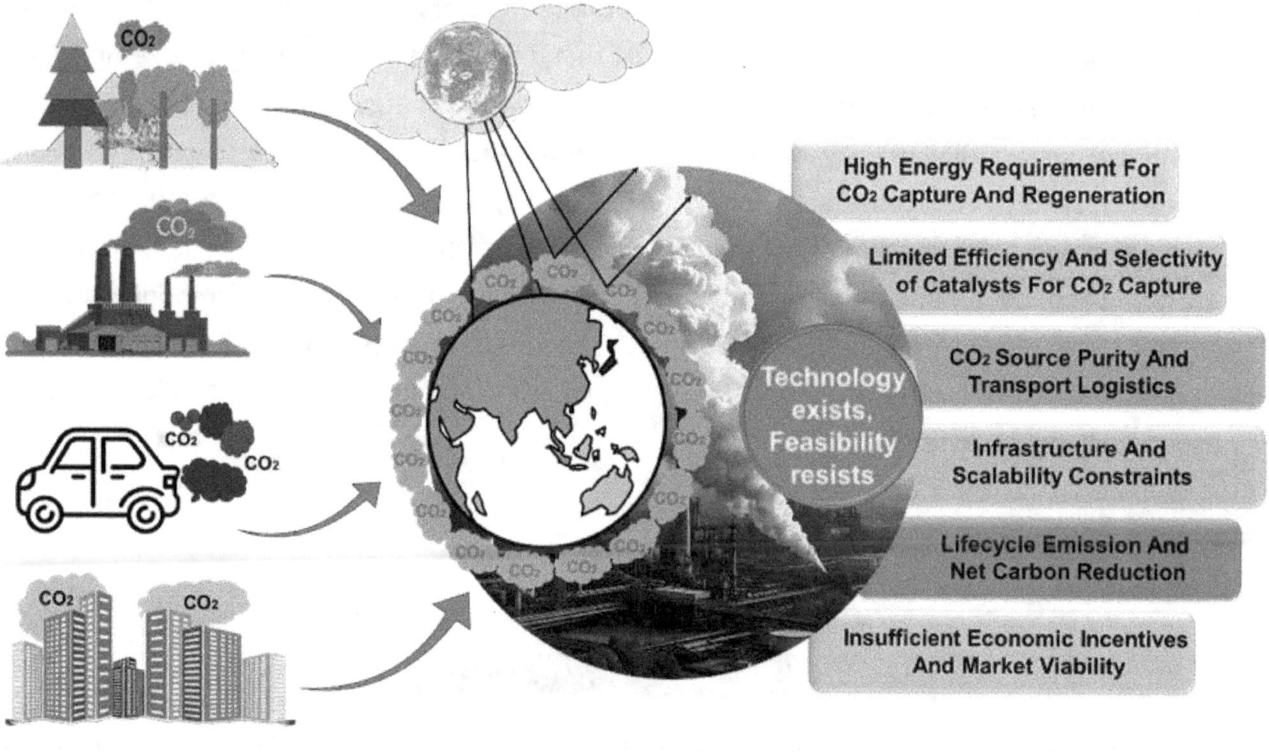

Figure 3.1: Different sources contributing to the contribution of CO_2 and the technological domains for which progress is required to combat CO_2 production. Copyright Credit: (Ref. [13], Energy and Environmental Science, RSC, 2026)

All the mentioned are possible ways through which CO_2 emissions can be controlled. Among the mentioned possibilities, the first and the last ones are quite accessible, but demand efforts from the community level to the geopolitical levels for shifting the regime of energy [5]. The second option, although a major contributor to the CO_2 emission in the atmosphere, is a bit tedious and requires a lot of research work for maintaining the efficient output at the cost of less input. For the immediate startup and to begin seeing results in a short span, the third option is the most viable way opted by the research community for attaining effective results along with the attainment of the zero-emission target. For accomplishing the process of CCS, it has been subcategorised into different stages, that is, capturing processes: pre-, post-, and oxy-combustion, evaluation and implementation of natural gas sources, DAC, along with the assessment of other gases generated through industrial processing [6]. Besides that, the progress made in the field of utilising and implementing the renewable energy sources acts as an effective source of replacement for fossil fuels. But looking into the role of energy in every sector of life, fossil fuels will remain a significant source of energy until the transition is complete, which may take some time, given recent developments in renewable energy. Among the different stages, the primary one, which is the phenomenon of pre-combustion, results in the evolution of syngas as the fuel interacts with the steam, air, and oxygen. The generated gas is being processed by the water-shift reactor, which results in the development of CO_2 and H_2 gases. The two important obstacles that need to be resolved in this process are cost-effectiveness and a practical model for the large-scale production [5]. In the second process, the interaction between fuel and oxygen results in the production of CO_2 as a by-product. In this process, there is no need for any device for separating the CO_2 as it comes directly as a by-product. However, the challenge associated with this process is supplying an adequate amount of oxygen for the interaction with the fuel, which requires a high amount of energy, and its incomplete commercial readiness [4, 5]. The third process, which is post-combustion, is the one currently responsible for the major contribution of CO_2 emissions. The major sources of emission in this process are contributed by fuel (coal and natural gas), gas-based plants [7]. While assessing the composition of fuel gas, it has been estimated that for coal-based plants it is: 12–15% CO_2 with 5–7% H_2O, 3–4% O_2 and 74–80% N_2, while for the natural gas-based plants it is: 4–10% CO_2 with 5–10% H_2O, 3–8% O_2 and 72–88% N_2 [8]. Besides that, the output at elevated temperature has also been found to possess a considerable concentration of SO_x and NO_x. Still, the main interest and focus of study in the research community has remained focused on the understanding and developing the technology for the separation of CO_2 and N_2, without taking the other gases into consideration [9]. The capturing of CO_2 through the post-combustion process is easily retrofittable with the existing power plants, as it focuses on the final output stream of the plants [4]. The main hurdle resides in the large energy consumption for the output fluxes containing a low concentration of CO_2. Furthermore, the advancement of post-combustion carbon capture is impeded by the requirement for large volumes of pre-treatment units. It has been observed that retrofitting the capturing assemble with the power plants results in increasing the cost of the generated electricity by at least 35%, which is against the aim of making the energy production cost-effective. Therefore, it demands further attention for the progress so that the capturing of CO_2 will be an economically viable process [10]. Above all processes, DAC is a viable way for bringing negative emissions and thereby decreasing the surplus amount of CO_2 in the living atmosphere. But there are still two significant issues with this technology's potential for use in the future. The primary limitation is imposed through the fundamental rules of thermodynamics, which state that a low concentration of CO_2 prevents the sorbents' performance in terms of capturing. The second challenge is the high-energy requirement for the production and regeneration of sorbent [11]. Although there are challenges, every process makes progress with the advanced technological developments [12]. Different engineering techniques have been implemented to progress the field of carbon capturing. In this book chapter, a brief description of the gap between material development and engineering will be discussed. Then a brief description about some important capturing processing will be provided. The section will be followed by the different surface engineering techniques employed for material designs pertaining to capturing. Finally, some hybrid materials will also be discussed, followed by the challenges, future perspective and conclusion.

3.2 Importance of Bridging the Gap Between Material Development and Surface Engineering

Materials that are being used in carbon capturing in the form of adsorbents, membranes, etc., are very important for the capturing processes taking place through either adsorption or using membranes. Different materials are being employed for the capturing process in different ways [13]. However, the characteristic features: that is, purity, reversibility, as well as energy-efficient of the physical adsorption process using porous materials have made these materials attractive. These porous materials are the solids which hold micro to macro size voids or gaps. It is through these gaps/voids that these materials permit the passage of different gases, liquids or small particles [13]. It has been more than a decade now that the CO_2 physisorption has been studied using different porous materials, inorganic and organic in nature [14]. Among them, the materials which were found to show significant potential for CO_2 capturing are zeolites, carbon (activated), porous organic polymers and aromatic frameworks (POPs & PAFs; **Figure 3.2 (a)**), some metal and covalent -organic frameworks like MOFs & COFs **(Figure 3.2 (b))** [15]. Although all these materials have showed the capability at the lab scale, but for bringing these materials to industrial scale lot of challenges comes into play that are cost-effectiveness of the synthesis procedures, extended stabilities, and geometrical configurations. However, different aspects of these have been explored, such as surface area, structure, pore size, interaction nature, stability, and morphology [2, 16]. But for the successful transition from the lab-scale to the industrial scale, it is necessary that the materials undergo engineering before proceeding with the industrial implementations. As an example, although the materials Mg-MOF-74 possess a large capacity for CO_2 capturing, it has been found that it requires high-energy inputs for blowdown and evacuation operations while processing [17]. Thus, further exploration is required before proceeding to the practical implications at the industrial level. Although continuous efforts are being made, some of the prominent aspects that need in-depth analysis are: how a material will bring impact on the current regime of CO_2 capturing technology, how the material can be synthesised in a cost-effective approach, and how to establish the potential of the material before proceeding to the market level. These are the questions that one needs to search for and can prove to be the best way that can be done through engineering. In this way, the gap can be bridged, and thereby it can act as a holistic approach for attaining the goal of zero-carbon emission.

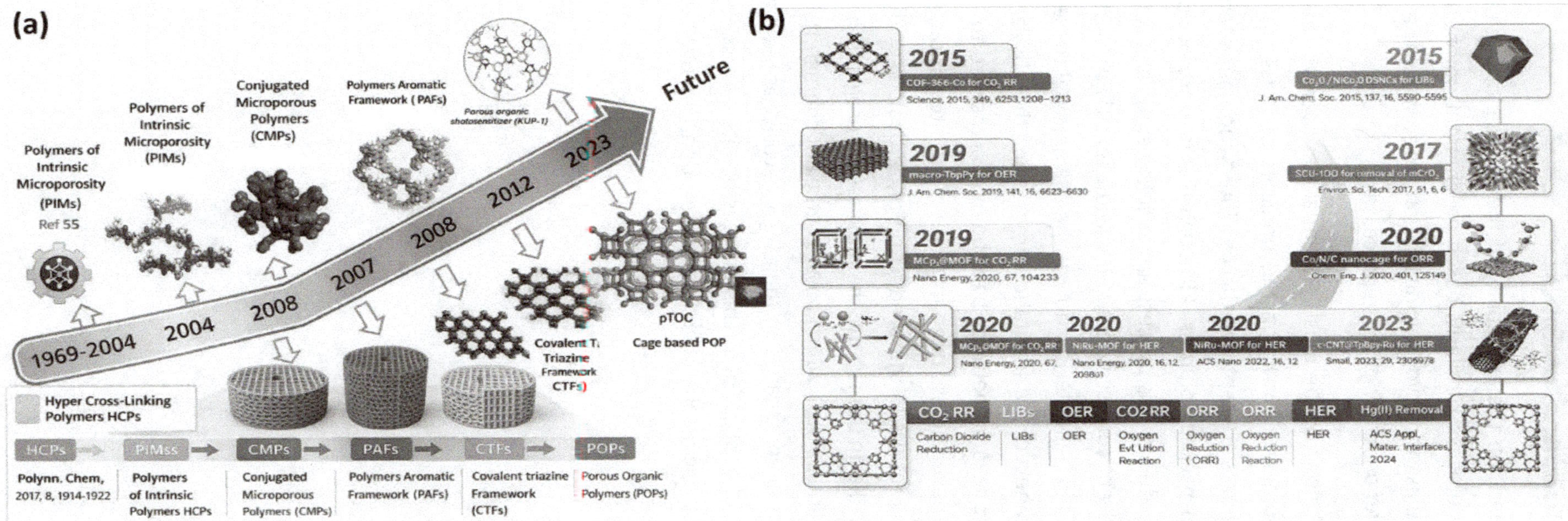

Figure 3.2: (a) A schematic overview representing the progress and significant contribution of amorphous POPs over time, (b) Schematic timeline of the success achieved in typical COFs and MOFs, along with their applications in different energy domains. (Copyright Credit: Ref [18], Materials Today Sustainability, Elsevier, 2024 , and Ref [15], The Innovation, Elsevier, 2025)

3.3 Process Employed for Carbon Capture

The design of any process depends on the availability of the sources or the raw materials, which are the basic building blocks of any process. Similarly, based on different point sources, the processes of absorption, adsorption, membranes, and cryogenic distillation are some of the processes employed for carbon capture. Let us have a brief overview of these processes:

Absorption: This technique is one of the most influential and sophisticated techniques employed for CO_2 separations. This phenomenon works on the principle of mass transfer between different phases. In this process, the CO_2 fluxes are captured through interaction between the gaseous and liquid phases occurring in a solvent of packed columns. Although this is a profound process for capturing, some of the main challenges/hurdles in this process are as follows: (i) requirement of high energy for the regeneration process, (ii) solvent corrosiveness, (iii) solvent sensitivity to different flue gases apart from CO_2, such as SO_x and NO_x, and (iv) deterioration of the solvents. All these points are the primary reasons which lower the effectiveness of the separation process and thereby raise the cost of the power generation. Despite the fact that several methods have been documented to lower the energy needs in the absorption process, they nevertheless need intricate plant integration. Besides that, the usage of amine solvents in this process results in the production of aerosols as well as nitrosamines as by-products, which are hazardous to life. In this regard, the most recent progress and thereby solution were put forward in the form of Ionic liquids (ILs), which possess efficient characteristic properties. However, their long-term utilisation is restrained because of their high viscosity as well as low capacity, thereby rendering their usage for practical applications. This process has its own subtypes depending on the approach of the absorption called chemical and physical absorption. The overview of both processes is depicted in **Figure 3.3** [19].

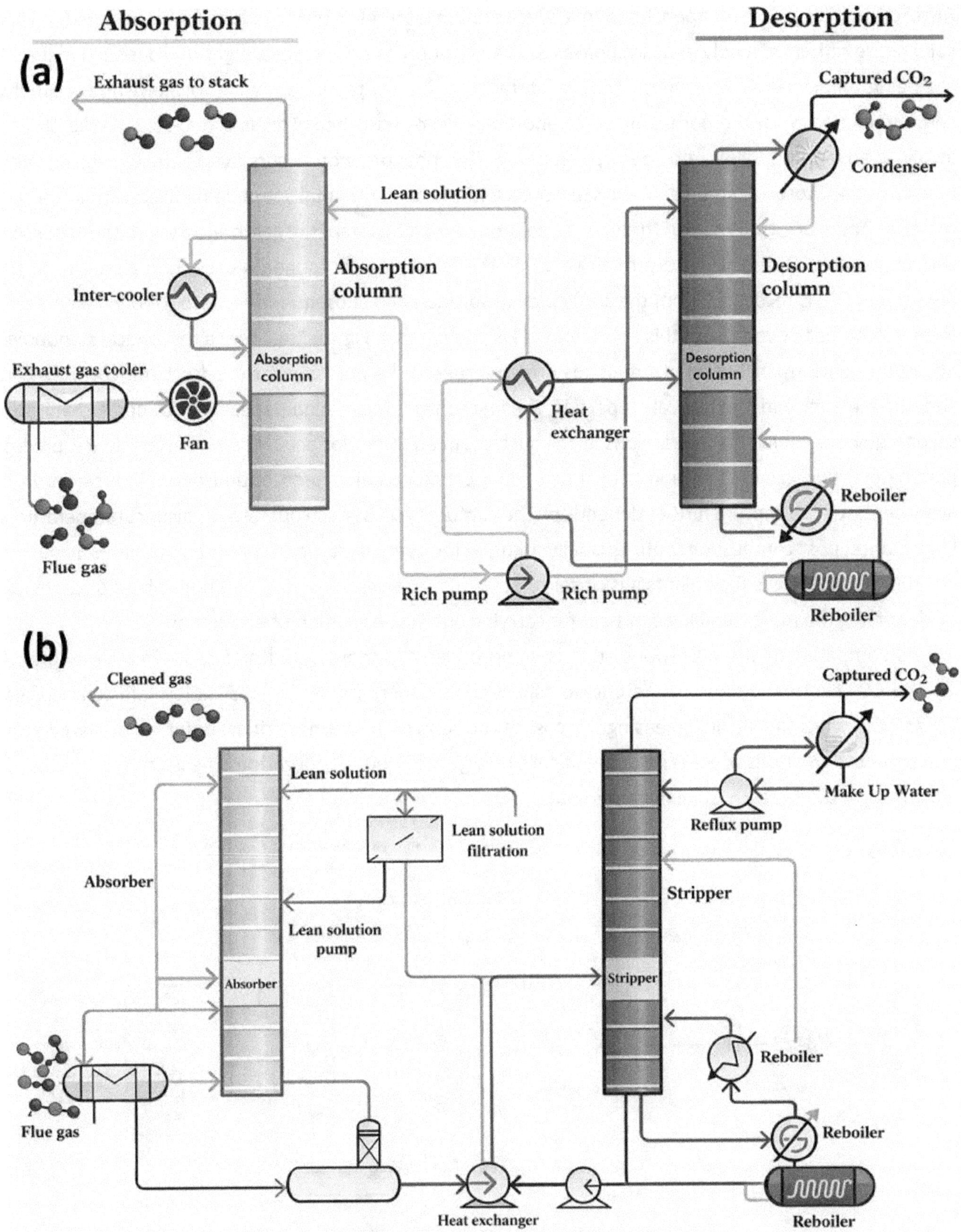

Figure 3.3: Schematic flow diagram for the (a) Chemical absorption process. (b) Physical absorption process is employed for the CO_2 separation technology. Copyright Credit: (Ref. [19], Fuel, Elsevier, 2023)

Adsorption: This process is opposite to that of absorption, as it relies on mass transfer among the fluid and solid phase rather than between two phases as in absorption. In this process, highly porous solid materials are being employed as adsorbents. During interaction, the gas particles, owing to their surface affinity, come in contact with the porous particles and thereby get adsorbed. This is a two-stage cyclic process involving adsorption followed by desorption of CO_2. The phenomenon occurs due to the uneven sorption affinity of the chosen adsorbent. The cycle begins with the experimental setup being filled with adsorbent material. Then, CO_2 gas is passed through the column. During this step, the gas interacts with the adsorbent surface, where CO_2 molecules are retained while the remaining components pass through easily. In the next stage, CO_2 is recovered from the surface by desorption and subsequently transported in compressed form. A schematic overview of this entire process is presented in **Figure 3.4**. From a theoretical standpoint, adsorption performance primarily depends on two aspects: the porosity of the adsorbent, covering pore size, surface area, and its chemical properties. The first controls the available space for CO_2 uptake, while the second governs molecular interactions at the surface. In addition, for achieving higher efficiency, process parameters, flue gas composition, and surrounding conditions also need optimisation. The selection of adsorbents for CO_2 capture further depends on the capture type, as each operates at distinct temperatures. For instance, post-combustion capture usually occurs at low temperatures; in coal-fired flue gases, it is about 60–100°C, while in gas-fired plants it is closer to 100°C. Thus, environmental conditions play a decisive role in identifying the most suitable adsorbent for carrying out the adsorption phenomenon [19].

As compared to the absorption, this is a more viable, sophisticated and cost-effective approach for the CO_2 capture. However, the critical aspect that hinders its progress is its low efficiency of gas separation, which results in increasing the cost of the supply [20]. Another critical point for the large-scale implementation of this process is the significant energy required for operation. These are the hurdles on which currently research is being carried out.

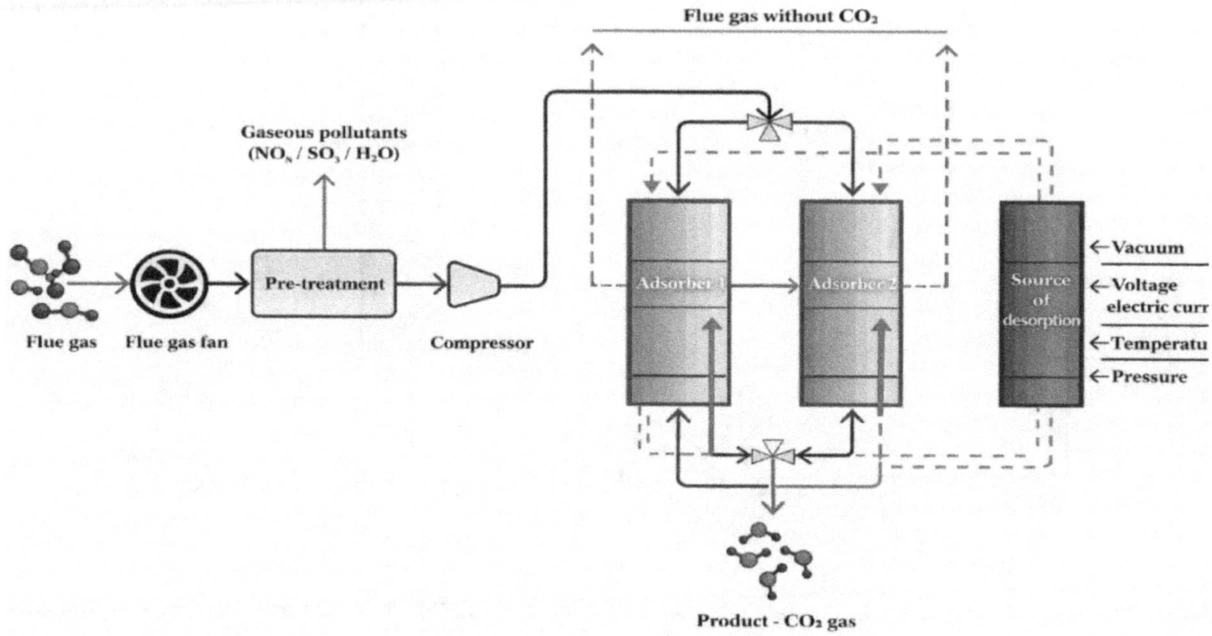

Figure 3.4: Schematics depicting the overview of capturing CO_2 through the adsorption process. Copyright Credit: (Ref. [19], Fuel, Elsevier, 2023)

Membranes: The capturing through membrane-based technology is considered to be one of the energy-effective methods, which even utilises less energy than that of the traditional ways of separation. This process can operate easily in a steady-state way without any requirement for high power for functioning. These membranes work on the selectivity process, through which CO_2 extraction is being done for further processing. Based on selectivity, the membranes allow the CO_2 to pass through, while restricting the other constituents present in the gas to permeate. Besides the selectivity, the parameters that govern the efficiency of the membrane are the pore size, volume, and permeability of the medium. Although this is currently one of the prominent methods frequently used for the capturing process, the main challenge for the membrane-based systems is to develop a system with optimised parameters necessary for capturing [21].

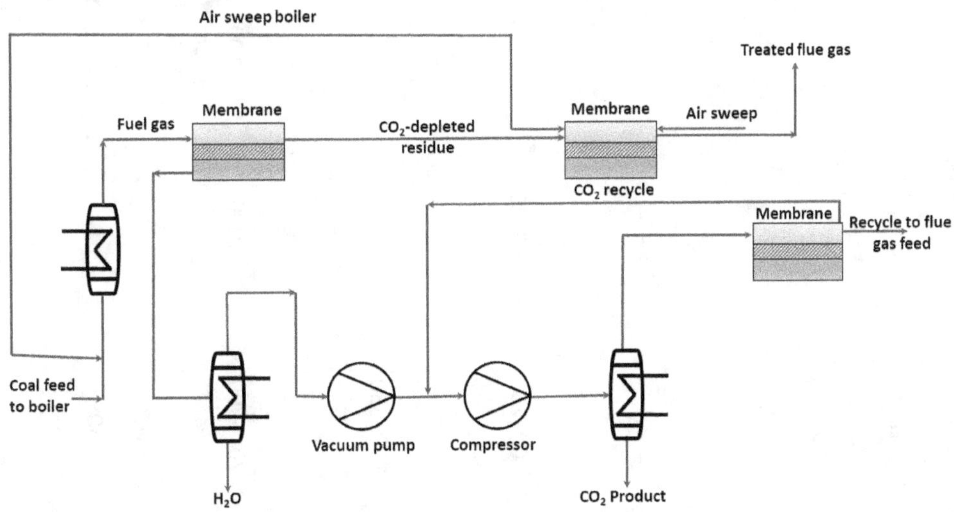

Figure 3.5: Simplified engineering of the post-combustion-based capturing of CO_2 from the flue gas emerging from the power station. Copyright Credit: (Ref. [19], Fuel, Elsevier, 2023)

Regarding the classification, the membranes are categorised based on their origin, morphology, and structure. Based on the materials, industries usually favour the usage of the organic-based membranes. The organic-based membranes have a variety in terms of their physical structure as well as the available forms of materials. They offer excellent stability even under high-temperature, pressure, and reactive environments. Looking into these aspects, the well-known class of membranes that are polymer as well as their hybrid systems are being considered to be very effective for the CO_2 separation, owing to their excellent permeability, selectivity, efficiency, as well as ease of regulating the pore size while synthesising them. On behalf of their structure, they are being classified into four types: porous, homogeneous solids, membranes with electric charge, solid or liquids with selective carriers. Besides these, the structure of membranes can be symmetric or asymmetric, that is, either there will be no change across the cross-section, or else there will be a change across the cross-section. Different projects related to the membrane-based technology have been established from 2022 onwards; the information about some of these are tabulated and is presented in **Table 3.1**. Also, a schematic overview of the post-combustion based on the flue gases from the power station is depicted below in **Figure 3.5**.

Besides the above-discussed strategies, some other ones which are in play for capturing technology are chemical processes, along with cryogenic methods, for the separation.

Table 3.1: Different membrane-based projects for CO_2 capturing, along with their various parameters across the globe.

S.No.	Type of membrane	Project	Prime performer	CO_2 pressure normalised flux [GPU]	Operating temperature [°C]	Manufacturing cost for membrane material [\$/m²]	Module cost of manufacturing and installation	Cost of CO_2 capture [\$/tCO_2]	CO_2 Recovery [% vol.]	CO_2 purity [%]
01	Carbon molecular sieve hollow-fibre membranes	Capturing CO_2 from coal-syngas	State University of New York	200	150	20	~1000 \$/kg/h	–	90	95
02	Polymeric membranes	CO_2 capture in IGCC power plant	Membrane Technology and Research, Inc.	225	200	500	15 \$/kg/h	84	90	99.5
03	Amine-containing polymeric membranes	CO_2 capture from coal-derived syngas	Ohio State University	327	107	~54	97 \$/m²	–	90	95
04	Polybenzimidazole polymer hollow-fibre membrane	CO_2 capture from a syngas stream	SRI International	80–120	200–250	30–80	–	–	–	–
05	WGSR catalytic membrane reactor	Integration of WGSR, H_2 separation, and CO_2 enrichment	Bettergy Corporation	150	350–550	1200	–	–	80	95
06	Ceramic-carbonate dual-phase membrane reactor	Integration of WGSR, H_2 separation, and CO_2 enrichment	Arizona State University	300–600	700–900	1000	~1000 \$/kg/h	–	99	90
07	Polymeric membrane	Large pilot polymer membrane system	Membrane Technology and Research, Inc.	1000	30	–	50–100 \$/m²	–	70–75	99
08	Mixed-matrix membranes (rubbery polymers, metal-organic polyhedral)	Achieve high CO_2 permeance	State University of New York	1500–2000	60	–	–	30	–	95

09	Polymeric composite membrane	CO_2 capture from flue gas	The Ohio State University	3500	57–77	20	40 \$/m²	40.0–41.5	60–90	95
10	Polymeric membrane	Retrofit process of polymeric membrane CO_2 capture system	Membrane Technology and Research, Inc.	1000	30	10	50 \$/m²	57.64	90	96
11	Polymeric membrane	CO_2 capture from coal flue gas	Membrane Technology and Research, Inc.	1700	30	50	–	–	75	85
12	Molten hydroxide dual-phase membrane	Membrane support materials (metal oxides)	Luna Innovations	800	300	300	–	–	99	96
13	Amine carriers in nanoporous polyethersulfone support	Selective membranes for <1% CO_2 sources	Ohio State University	2299	57–67	10	32 \$/m²	246	90	95
14	Polyimide-based membrane	Combine with cryogenic separation to reduce CO_2 capture cost.	American Air Liquide, Inc.	–	−30 to −45	–	100 \$/kg/h	–	80–90	58
15	Graphene oxide-based membrane	Retrofit high-selectivity membranes in coal/gas power plant.	Gas Technology Institute (GTI)	1020	80	–	–	≤40	70–90	95
16	Graphene oxide-based membrane	Retrofit high-flux membranes in coal/gas power plant	Gas Technology Institute (GTI)	2500	70	–	–	≤40	70–90	95

3.4 Different Strategies for Surface Engineering

The phenomenon of surface engineering is an important aspect which plays a crucial role in the advancement of CO_2 capturing technology. It tends to modify the materials by changing their adsorption capacity, selectivity, stability and other characteristic property-defining parameters. Through these approaches, it has become possible to tailor the porosity, morphology, and chemical functionality of the materials, which not only helps in effectively trapping CO_2 but also plays a significant role in increasing the stability of materials over multiple cycles. The different ways that have been employed for carrying out the phenomenon are functionalisation, heteroatom doping and the development of hybrid materials [22]. These strategies tend to enhance both the chemical affinity and the stability of the materials employed for the capturing process. Also, regulation of the pore size and structure so that it matches well with that of the CO_2 molecule helps in increasing the efficiency even at low pressure [23]. The incorporation of meso- and macropores boosts the gas diffusion process in the materials. The surface modification process has also benefited the membrane-based CO_2 capturing technology. The surface modification through the coatings, such as amine grafting, MOFs and COFs, results in improving the selectivity and efficiency of the materials even under humid atmospheric conditions [24]. Thus, the different strategies applied for the surface engineering helped a lot in the progress and development of high-performance and cost-effective materials for the capturing process [25]. A brief outline about few of the important ones is provided below:

Carbonation (Mineral): Presently, the carbonation of the minerals seems to be the only practical applicable solution through which CO_2 can be stored. This process is also known as mineral sequestration and was proposed by Seifritz in 1990 [26]. There exist plenty of naturally occurring minerals as well as industrial wastes which can be employed for the sequestration/carbonation process for the CO_2. It has been estimated that among the total CO_2 emission sources, at least 10–15%, emission sources are those which possess potential for the mineral carbonation process [26–31]. Across the geological time periods, it has been observed that the silicate rocks undergo slow breakdown. While analysing, the phenomenon was attributed to mineral carbonation, which is the interaction of metal oxides with $CO_{2,}$ resulting in the formation of insoluble solid carbonates. On the broader aspect, the phenomenon of mineral carbonation is being categorised into two distinct types, that is, in-situ and ex-situ. In the primary process, CO_2 is being injected into the geological deposit for the formation of carbonates. However, the latter process is being carried out from the already mined rock over the ground surface [32–35]. While assessing the energy states for both states, that is, carbonates and CO_2, it has been observed that the primary one possesses an energy state in the range of 60–180 kJ/mol, which is quite lower than the latter one with an energy state value of 400 kJ/mol, as depicted in **Figure 3.6 (a)** [36]. Owing to the eco-friendly nature and stability of the carbonates, the phenomenon of carbonation is considered a long-term and leak-free approach for CO_2 disposal [37]. The ex-situ approach for the mineral carbonation can be carried out through direct and indirect approaches [38]. Among the two approaches, the most straightforward is the direct approach method.

The process of CO_2 mitigation through mineral carbonation has received great attention in the research community and is currently being investigated on a much larger scale. The phenomenon of direct is usually accomplished through direct gas- or aqueous mineral carbonation [38]. Among the two, the primary one is considered as one of the viable and basic approaches for direct-mineral carbonation, where CO_2 reacts with minerals and generates carbonates as by-products [39, 40]. The latter one is a single-step carbonation process, in which the interaction between CO_2 and minerals takes place in the aqueous medium [41, 42]. The different approaches involved in the ex-situ mineral carbonation, either through direct or indirect means is schematically shown in **Figure 3.6 (b)** [38].

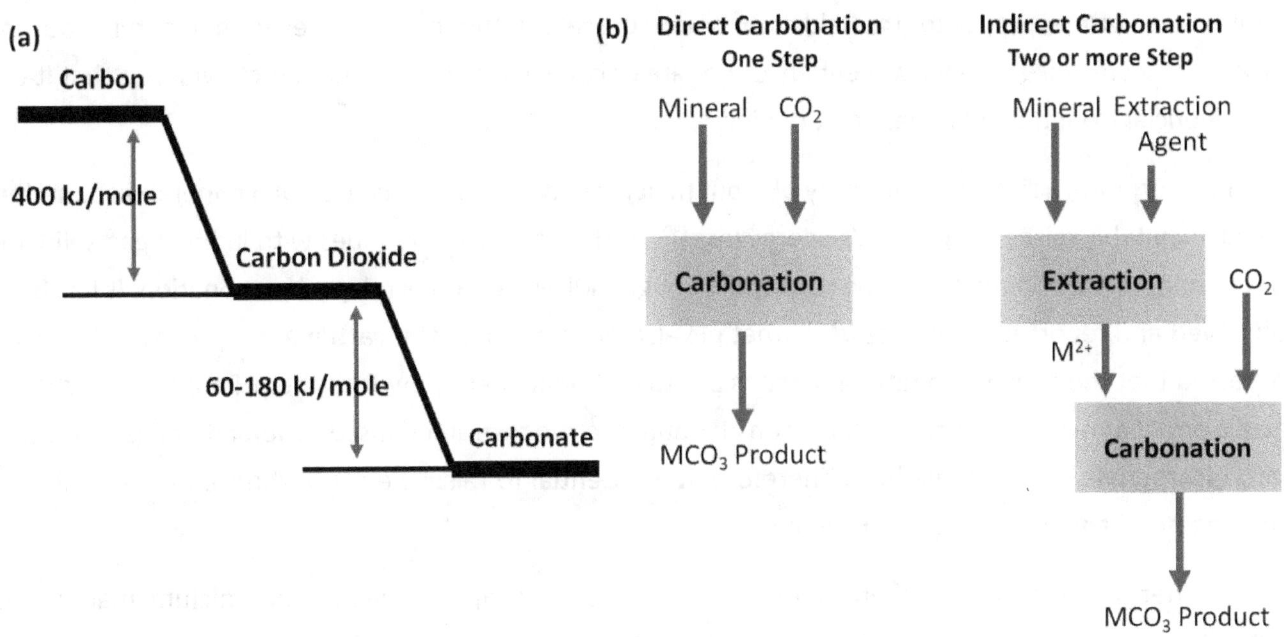

Figure 3.6: (a) Energy structure of different carbon phases. (b) Schematic overview of the direct and indirect carbonisation.

On the other hand, when the carbonation process occurs in multiple steps, it is called the indirect carbonation process. There are different approaches through which this indirect carbonation can be accomplished, which involve: pH swing and molten salt processes, HCl extraction, bioleaching, ammonia, caustics and different acidic extraction approaches [38, 43]. In addition, some of the popular feedstocks include both natural minerals and industrial wastes. **Figure 3.7** summarises the direct and indirect mineral carbonation techniques that are currently accessible in the literature. In recent research studies, it has been observed that this phenomenon is effective even at the intermediate stage: calcium looping. Under the CO_2 atmosphere, the successful gas-solid carbonation interaction takes place at a temperature of 650°C. However, for the reverse process, that is conversion of carbonated to the respective oxides take place at a temperature above 650°C, under a nitrogen atmosphere [30, 44]. The main drawback that hinders the industrial scale-up of the process is the sluggish rate of gas-solid interaction. While comparing the carbonation rates of the gas-solid and aqueous mineral, it has been observed that the latter process possesses a higher rate due to the utilisation of NaOH, NaCl, $KHCO_3$, and other substances [27, 38]. Thus, for attaining the required and necessary convergence rates, the procedure still has to be modified to some extent. There are certain limitations associated with the carbonation process, that is, mineral and CO_2 dissolution, control over CO_2 diffusion, owing to which its commercial progress is still unviable [27, 34, 38]. Among the discussed approaches of carbonation, the indirect approach is preferred over the direct as it yields the products with much purity [34, 38]. Besides that, in the indirect approach, it has been observed that the rate of conversion for calcium and magnesium to be transformed into carbonate is much higher than that of the direct approach. For the address, the issue pertaining to the sluggish response rate of the direct gas-solid mineral carbonation, the process was modified by employing the indirect stepwise approach for the said carbonation process. Although it is believed that carrying out the process in a stepwise manner will enhance the rate of conversion, there is no direct evidence that it can be beneficial from the industrial point of view.

Furthermore, a significant quantity of input heat is needed for indirect gas-solid mineral carbonation, which inevitably raises the process's energy loss [30, 38]. Thus, the current issues with indirect gas-solid and direct-mineral carbonations may be resolved through indirect aqueous mineral carbonation. It has been observed and reported by Park *et al.* [45] that pH also plays a role in the carbonation process. They have reported that the conversion rate of magnesium to carbonate gets slower as the pH moves towards the acidic environment. This decrease has been attributed to the restrained dissociation rate of carbonic acid into bicarbonate or carbonate ions. Therefore, it is essential to raise the pH of the solution to enhance the degree of aqueous mineral carbonation.

Furthermore, it has been observed that carbonate precipitation and the calcium magnesium extraction process are favoured in completely opposite pH conditions. The primary one is favoured in the high pH atmosphere, while the latter one is favoured in the lower pH atmosphere. The precipitation of calcium and magnesium carbonates is crucial, even though the employed acids tend to remove substantial amounts of their ion from the feedstock. Thus, raising the pH of the solution to about 10 appears to be a useful method for developing a process which can yield the carbonates with high purity [46]. Therefore, the secret in creating high-purity goods with high carbonation rates is a pH swing process. This process takes place in two stages: the first stage, which takes place in an acidic environment, involves the extraction of calcium and magnesium ions, and in the second stage, which takes place in a high pH environment, is the precipitation of the carbonates [46].

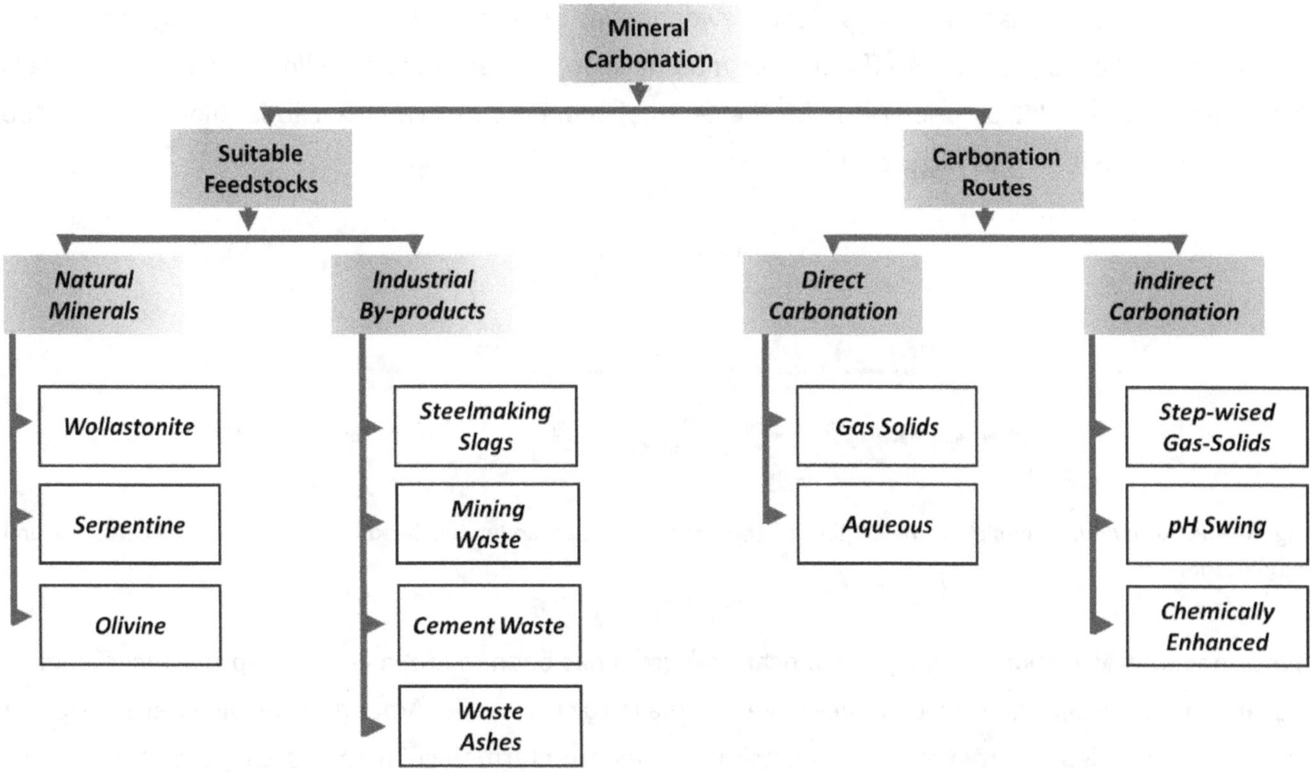

Figure 3.7: Schematic representation of the mineral carbonation for the CO_2 capturing and storage, with the subclasses in suitable feedstocks and carbonation routes.

Textural modifications: So far, different efforts have been made by the materials scientists and implemented by the research community for developing the sorbent materials with high-performance capacity. The CO_2 capture efficiency of the adsorbent material relies on many factors, among which textural properties are of particular importance. The textural properties, such as heat capacity and thermal conductivity, play a significant role in the CO_2 capturing efficiency of adsorbents [47]. The cyclic adsorption is an intricate process which involves many steps, such as adsorption, desorption, pressurisation, depressurisation, and purging. Each step involved in the process has its thermodynamic, kinetics and dynamic aspects, thereby making them very intricate in nature. Due to this complex nature, it has become too expensive to carry out experiments for gaining insights related to the process parameters, simulation factors, and optimisation. Among the different parameters of modelling and design, such as number of cycles and time, pressure and flow rate, temperature, and bed configurations, the adsorbents' temperature control is an important parameter. The regenerative energy for the sorbent is defined as the total sum of the energy required for raising the sorbent to its regeneration temperature, energy required in the CO_2 desorption process, and the latent heat of vaporisation corresponding to the water molecules in the sorbent. The term regeneration temperature is defined as the product of the heat capacity and temperature difference. The latent heat of vaporisation is governed by the amount of moisture in the adsorbent, regenerative temperature, and time required for completion of the process [48]. Sorbents which possess lower heat capacity generally require less supply of heat to acquire the desired regeneration temperature [4].

From the above discussion, it is clearly evident that textural modifications play a significant role in governing the CO_2 capturing efficiency of materials. To gain an understanding of how this impacts different materials, let us briefly discuss the textural properties of carbonaceous, biomass-activated carbon, and COF materials **(Figure 3.8)**.

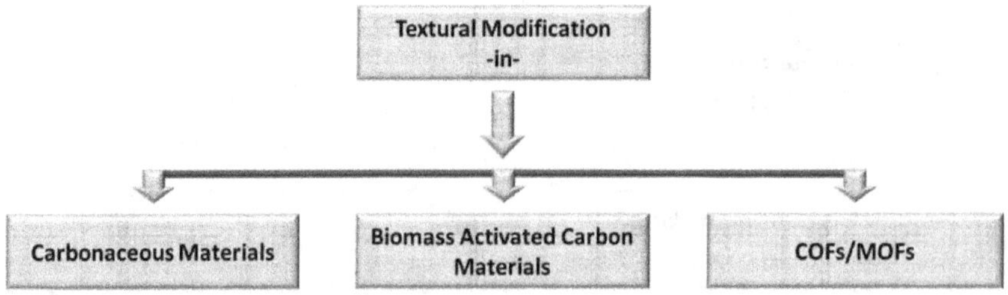

Figure 3.8: An overview depicting the outline of the section, a head under the heading of the importance of textural modifications.

Carbonaceous Materials: Although significant progress has been made in carbon capture technologies, advancements in separation techniques have also played a crucial role. Among these, physical adsorption stands out as a leading method. This technique relies on intermolecular forces and is preferred over others due to its ease of regeneration. Adsorption methods are particularly favoured for CO_2 capture because they are cost-effective, energy-efficient, and versatile, functioning well across a wide range of temperatures and pressures [49]. The choice of adsorbent material is based on several key factors, including its adsorption capacity, selectivity, stability, and regeneration potential. To be effective, a CO_2 adsorbent should have a high adsorption rate, strong CO_2 uptake, excellent selectivity, the ability to be reused multiple times, and robust thermal and mechanical stability [50]. A variety of materials have been explored for CO_2 capture, such as activated carbon, carbon fibres, zeolites, carbon nanotubes, graphene-based materials, MOFs, COFs, and metal oxides [51, 52]. Among these, various carbon-based adsorbents, developed through different modifications **(Figure 3.9)** [53], have demonstrated outstanding CO_2 adsorption performance under diverse environmental conditions.

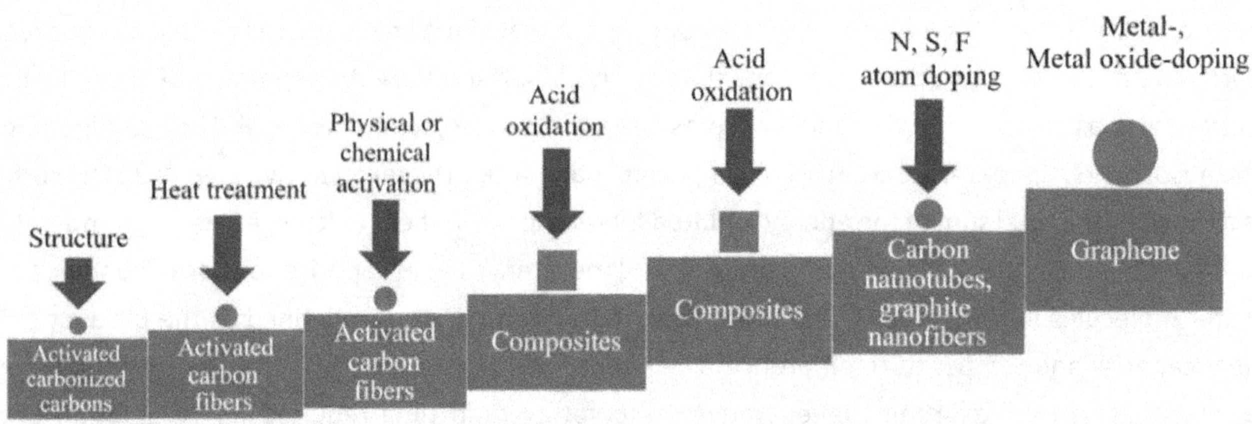

Figure 3.9: Surface engineering progress of the carbonaceous materials over the course of time. Copyright Credit: (Ref. [53], Journal of the Taiwan Institute of Chemical Engineers, Elsevier, 2017)

In this section, among the different surface modifications that carbon-based adsorbents have undergone, a brief discussion about the textural pore structure will be presented. Among the various factors, pore structure and the surface area play a crucial role in the adsorption capacity of a material. In this regard, as per the studies reported by Presser *et al.* [54], it is very difficult to generate a clear relation between the pore structure and the amount of CO_2 adsorbed. While analysing, they have observed that although there exists a linear relationship between the BET surface area and CO_2 adsorption capacity, the portion with higher surface area does not correspond to the maximum CO_2 uptake capacity. It has been found that the involvement of pores with different diameters in the adsorption depends on the pressure. Pores with a diameter of the order of 0.5 nm and less were found to contribute to the process mainly at low pressure, whereas pores with a diameter of 0.8 nm have been found to contribute to the process at high pressure. Similarly, Sevilla *et al.* [50], in their analysis of the CO_2 adsorption capacity of the porous carbon materials, have found that the contribution is mainly from micropores with large volume, rather than from surface area, pore volume, or the heteroatom. Pertaining to the pore size, they have also observed a similar behaviour reported by Presser *et al.* [54], and thus came to the conclusion that it is actually the volume corresponding to the ultramicroscopic pore which determines the efficiency of CO_2 capturing capability of a material from flue gases. When taking into account the cyclability, it has been observed that activated carbon-based materials exhibit excellent performance. Similar to the porous carbon materials, the adsorption capability of the activated carbon materials is also dependent on the volume of the ultra-micropores for capturing CO_2 from flue gases. As anticipated, given the somewhat heterogeneous nature of an activated carbon's interior surfaces, the isosteric enthalpy of adsorption was higher at lower CO_2 loadings [55]. The activation by different means tends to improve the adsorption of these materials. In this regard, it has been observed that the direct activation of phenolic resin spheres through KOH by employing the modified Stöber technique results in the development of modified carbon material [56].

The material has been found to exhibit efficient CO_2 adsorption capabilities, attributed to the presence of a substantial proportion of micropores below 0.8 nm. The CO_2 uptake capacities were noted at two different temperatures, that is, 0°C and 23°C, and the recorded values were found to be 8.9 mmol/g and 4.6 mmol/g, respectively [56]. Although as such, no correlation was found for the pore volume and BET area with the CO_2 adsorption capacity, a linear correlation was evident between the micropores' surface area and the CO_2 adsorption capacity. Among the different available pores, it has been found that those with a size less than one nanometre are significantly contributing to the CO_2 adsorption process [57]. Although CO_2 possesses a linear structure with no molecular polarity, the bonding arrangement still plays a role in the CO_2 adsorption process. This effect has been verified by Wang *et al.* [57] through a theoretical simulation. In their study, they have found that for a nanopore with a size of 0.8 nm, if the CO_2 molecule lies at the centre, then the molecule will be linear in structure due to symmetrical forces from two attached O-atoms.

On the other hand, if it resides at the edge of the pore, then the CO_2 attains polarity due to London dispersion forces among the attached O-atoms and the C-atoms present in the active sites. This interaction leads to the generation of four different kinds of adsorption processes in the molecular CO_2, as depicted in **Figure 3.10 (a)**. Among the depicted configurations, the second and fourth ones possess both polarities as well as the dipole moment.

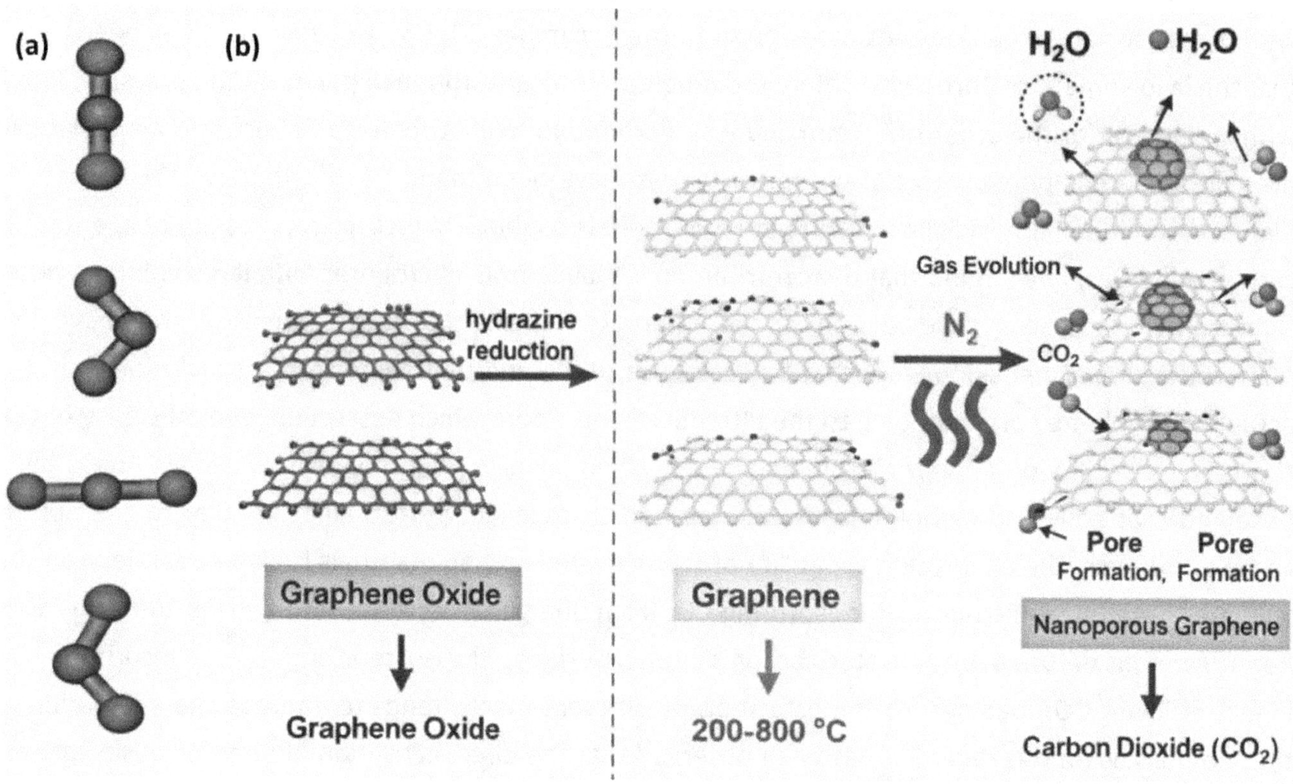

Figure 3.10: (a) Different CO_2 structures possible for the CO_2 adsorption. (b) Schematic overview of the pore evolutions in graphene nanosheets during thermal process in a nitrogen atmosphere. Copyright Credit: (Ref. [57], Chemical Engineering Science, Elsevier, 2011 and Ref. [58], Journal CO2 Utilization, Elsevier 2016)

Thus, the polar CO_2 molecules will get adsorbed to the active sites through chemical interactions, resulting in the formation of C-O complexes on the surface. Besides this, the London dispersion forces in the polarised CO_2 molecules tend to enlarge the pore size, which in turn results in the enhancement of the CO_2 adsorption and selectivity of the material. Graphene, another form of carbon, when sintered at 800°C for 2 hours in a nitrogen atmosphere, was found to exhibit excellent properties favourable for CO_2 adsorption. The material surface area and pore volume were found to be 484 m^2/g and 0.68 cm^3/g, respectively. Due to the narrow pore distribution, it was found to have a 2.89 mmol/g capacity for CO_2 capturing. At room temperature and 1 bar pressure, the materials present rapid kinetic separation and excellent adsorption capabilities for CO_2 and N_2. The evolution of the pores in graphene nanosheets during the thermal process as a consequence of gas evolution is shown in **Figure 3.10 (b)** [58]. In some of the recent studies, it has been observed that carbon materials which possess high selectivity with respect to CO_2/N_2 usually exhibit low CO_2 adsorption capabilities [59]. Thus, the isosteric nature of the carbon related to the enthalpy of adsorption has been attributed to the presence of a large number of surface functional groups. The thermodynamic analysis revealed that the adsorption of CO_2 by the carbon and activated materials is physical, exothermic, and spontaneous in nature. However, the isosteric nature of enthalpy as well as Gibbs free energy of the carbon materials shows a significant decrease with the rise in the surface loading [60]. There are a lot of reports following a similar trend as discussed, and progress is ongoing. However, there are some other approaches for the surface engineering of the carbonaceous materials. These approaches are mainly the alkaline activation, heteroatom doping, which involve modification of amine group, metal-oxide group, O-, S-, F-, etc. For a general outlook, **Table 3.2** provides an enlightenment about the different materials, activators, and the related achieved properties.

Table 3.2: Different characteristic features of modified carbon materials for CO_2 adsorption.

S.No.	Material	Activator				Surface area (m^2/g)	Micropore volume (cm^3/g)	Adsorption temperature (°C)	CO_2 adsorption capacity (mmol/g)	CO_2/N_2 selectivity	Ref.
		Agent	Alkaline: C (wt%)	Temp. (°C)	Time (h)						
01	Carbonised waste biomass	CO_2	–	800	0, 2	841	0.281	0	1.45	14.5	[55]
		H_3PO_4	–	600	1	1073	0.101	0	0.80	26.7	
02	Phenolic resin-based sphere	KOH	4:1	700	1	2400	–	0	8.9	–	[56]
03	Coconut shell activated carbon	NaOH (32%)	–	RT	3	378.23	–	35	0.62	–	[61]
04	Activated carbon monoliths	KOH	1:1	800	1	1665.7	0.73	25	3.202	–	[62]
05	Activated carbon	KOH	4:1	700	0.17	1636	–	0	7.49	98	[63]
06	Pitch-activated carbon	CO_2	–	900	2	527	0.26	30	0.56	–	[57]
07	Rice husk char	KOH	1:1	710	1	1041	0.42	0	5.63	–	[64]
08	Graphene sheets	N_2	–	800	2	484	0.094	0	2.894	112	[58]
09	Graphite nanofibers	KOH	4:1	900	2	567	0.274	25	1.35	–	[65]
10	Activated carbon fibres	KOH	3:1	900	1	2180	0.928	25	5.68	–	[66]

3.5 Biomass-Activated Carbon Materials (BAMC)

These adsorbent materials are among the recent materials developed from biomass and have gained interest because of their non-toxicity, accessibility, cost-effectiveness for disposal, and thermal stability [67]. Although ionic strength, temperature, pH, etc., are some other factors that can modify the composition of BACMs. and, in turn, their properties. However, focusing on the CO_2 adsorption capabilities of these materials, which are governed by the textural properties, including porosity, surface chemistry, and the surface area of the materials. It has been observed that the carbonisation process in the primary stages led to the removal of the volatile matter from the feedstock materials, which has a serious impact on the pore size of the material. In a study carried out by Serafin *et al.* [68], they reported that temperature has a direct impact on the surface area of the material. Besides that, they have reported that the metal-impregnated BAMCs materials developed at temperatures 272 and 298 K have a relative variation in surface area by 10 times. The materials developed at higher temperatures with increased duration of pyrolysis were found to possess a surface area 10 times greater than the other one. This increase has been attributed to the release of volatile substances.

In corn stalk BAMCs and modified BAMCs materials impregnated with Fe and Cu were found to possess a surface area of the order 603.24 m^2/g and 1316 m^2/g, respectively [69, 70]. From the different results, it has been found that the surface area of BAMCs' materials relies on the employed biomass. On comparative surface area analysis, it has been found that peanut shell-based BAMCs have less surface area than that of rice straw-based BAMCs, which has been attributed to the presence of varying concentrations of cellulose and hemicellulose. Similarly, among the pine and nutshell-based, the surface area of pine is higher [71]. Similar behaviour was observed for the iron-activated carbon material obtained from three different sources, that is, Mangifera indica, Ficus religiosa, and Ficus benghalensis, at 600°C, which were found to have different surface areas as 63.9, 47.08, and 30.26 m^2/g, respectively [72]. All the observation clearly validates that the variation in the surface area is due to the presence of different cellulose, hemicellulose, as well as other volatile matter. The increased surface area directly tends to enhance the CO_2 adsorption capacity of the material.

Another surface parameter that determines the CO_2 adsorption capability of the materials is the pore size. As per the IUPAC norms, the pores are classified as: macro-sized, pores with a size of 50 nm or greater; meso-sized, these are the pores with a size between 2 and 50 nm, and less than that form the class of micro pores [73]. For the better adsorption capacity of BACM, the pore size of the materials must be at least greater than the size of the CO_2 molecule, which is 0.33 nm [74]. In this regard, Dissanayake *et al.* have studied two different materials with pore sizes of 0.5 and 0.8 nm and reported that the latter material with a higher pore size possesses a much better CO_2 adsorption capacity. The number of pores tends to increase due to the carbonisation process, which tends to enhance the material interaction [75]. Also, the activation of BACM has been found to affect the pore structure of these materials significantly. Due to activation via KOH, garlic peel-based BACM has been found to produce an efficient CO_2 adsorption capacity attributed to the generation of a large number of micro pores in the material. However, an appropriate amount of the activating agent and proper temperature are necessary, as the overdose of the activating agent and temperature can damage the microporosity of the structure [76]. Another aspect related to the metal-impregnated BACM materials is their surface chemistry, which relies on the presence of CHNOS elements in the feedstock. The nature of the material that is acidic, basic, or neutral is determined by the CHNOS present in the stock. The adsorbent's nature is usually determined by the attached functional group. For the acidic materials, the attached functional groups are carboxylic acids, anhydrides, lactones, lactols, and phenols; for the basic materials, the functional groups are quinone, chromene, pyrone, and nitrogen; and for the neutral materials, the groups are carbonyl and ether [77]. A lot of research work is being carried out in analysing the functional groups in both metal-impregnated and without impregnated BACM materials. As the surface analysis of these materials will provide information about the surface area, nature (acidic, basic, or neutral), etc., thereby about the adsorption capability of the material.

COFs: Due to the fact that the heat capacity information pertaining to the different adsorbent materials is absent, for assessing the different calculations, the heat capacity value of standard Zeolite 13X is utilised. In this regard, Danaci *et al.* [78] have studied the effect of the sensitivity on the physical characteristics and thereby the impact on the metric processes of the adsorbents. In their study, they have observed that if the value employed for the heat capacity departs significantly from the actual value, it tends to have a negative effect on the recovery and purity of CO_2, and also results in decreasing the operational lifespan of the adsorbent. Thus, it is evident that the performance of a material is strongly dependent on its heat capacity value. The same thing has been verified in another study, where it has been observed that the predicted performance of a material showed a significant difference against the standard and the experimental heat capacity values of a material [79]. Due to the dearth of experimental evaluation of the heat capacity value of the adsorbents, it has become a hurdle for evaluating the same through molecular dynamic simulations. Different approaches are being employed for evaluating the value; for the same, recently, Moosavi *et al.* [82] employed the machine learning tool for the evaluation of the heat capacity of COF material. They have found that these materials do not have any specific value of heat capacity; rather, a bunch of values exist which reside in a range of 0.4–1.1 $Jg^{-1}K^{-1}$. They carried out the analysis further by considering a particular constant value of 0.985 $Jg^{-1}K^{-1}$ from the range and evaluated the actual amount of heat requirements. The observed results surprisingly demonstrated that these materials (COFs) tend to lower the heat requirement by approximately 50%. Thus, in order to get precise information pertaining to the performance of the material, it is necessary to evaluate the experimental value for the heat capacity of the newly developed COF materials.

Moving to the next important parameter, that is the enthalpy of adsorption (ΔH_{ads}), which provides us information about the nature of interaction occurring between the CO_2 and the adsorbent, along with the amount of energy required for the desorption process. For the different types of adsorption, that is, physisorption and chemisorption, the values of ΔH_{ads} are approximately around –35 kJ/mol and

–50 kJ/mol, respectively [83]. The higher value of ΔH_{ads} signifies the higher CO_2 capturing capability of the material, but on the other hand, it also depicts the requirement of more energy for the adsorbent renewal. Thus, for balancing both the CO_2 capability as well as the amount of energy required for the renewal of the adsorbent, the value of ΔH_{ads} for a material being employed for CO_2 capture must lie within the range of –25 to –50 kJ/mol [83]. It has been observed that most of the COF materials have their ΔH_{ads} values residing within this limit. However, in cases where the value of ΔH_{ads} does not reside within the range, then the thermodynamic characteristics of the said COF materials need to be changed, which can be accomplished by either altering the pore size or attaching a functional group, which results in regulating the value of ΔH_{ads}. It has been observed that through the incorporation of varying quantities of carboxylic acid in H_2P-COFs, it not only regulates the value of ΔH_{ads} between –32.2 and –43.5 kJ/mol but also enhances the CO_2 uptake concentration from 31 to 76 mg/g at a regulating temperature of 298 K [84]. Similarly, Gui *et al.* [80] have employed the click reaction for modifying the 3D COF. Employing the reaction technique, they incorporated multiple azide compounds in BMTA materials. Through this process, it was found that the value of ΔH_{ads} for the said material gets modified from –19.5 kJ/mol to –29 kJ/mol, thereby residing within the range. Similarly, through the modification of the phenolic group employing the 4-phenylazobenzoyl- and 4-stilbenecarbonyl-chlorides, in different COF materials results in enhancing the uptake CO_2 capacity from 32 to 115 mg/g along with the increment of ΔH_{ads} from –30 to –43.4 kJ/mol at a temperature and pressure of 298 K and 1 bar, respectively [81].

Another important parameter that describes the capability of an adsorbent to conduct heat is the thermal conductivity of a material. It has been observed that the materials which possess low thermal conductivity exhibit significant variation in the temperature across the column. While analysing the flow of heat through different materials, it was observed that the heat transport through the porous adsorbent materials occurs through both radiation and convection. The phenomenon of thermal conductivity is governed by the pore size and the nature of the materials. In this regard, it has been observed that two-dimensional (2D) COF materials exhibit exceptionally larger thermal conductivity than the organic/porous materials [85]. Recently, in a study carried out by Ma *et al.* [86], they found that the material COF-300 possesses high thermal conductivity greater than 15 $Wm^{-1}K^{-1}$ in every direction of the surface. In there, they have observed the dependence of the thermal conductivity on the pore size as well as knots. They have found that by choosing the different linkers, knots and pore sizes, these materials can yield desired anisotropy as well as thermal conductivity. The thermal conductivity of a material is sensitive even to the gases being adsorbed. It has been observed in a study that the density of the gas being adsorbed has a significant impact on the thermal conductivity of the material. A gas with higher density has been found to increase the thermal conductivity across the plane in the orientation of laminar pores. The increase has been attributed to the colliding interaction of the gas molecules within the framework.

On the other hand, the thermal conductivity of a material within the plane (2D Sheet) has been observed to decrease monotonically [87]. However, because of the extra heat generated, the exothermic adsorption process might raise the temperature even more, which would restrict the capture performance. Besides the mentioned properties related to the thermal aspects of the material, which are tuned through different engineering techniques, textural features are important from an industrial point of view. For instance, the phenomenon of the electrical swing adsorption (ESA) becomes more effective in cases where the material possesses high electrical conductivity. In regard to ESA experimentation, the most frequently used adsorbents are activated carbon and zeolites [88]. COF materials exhibiting the features of being completely organic, conjugated, modular and porous make them excellent from the electrochemistry point of view. Related to the thermal conductivity of the COF materials, different studies have been carried out. In a study carried out by Evans *et al.* [89], they have analysed the continuous crystalline film of boronated ester-linked COFs. The synthesised materials were found to exhibit a dielectric constant of the order of 1.6 along with high thermal conductivity of the order of 1 $Wm^{-1}K^{-1}$. Similarly, a new COF material based on piperazine has been found to exhibit excellent electrical conductivity properties with a value of 2.72 and 12.7 cm^{-1} for the pellet and film-based sample, respectively [90]. The textural property has been found to have a significant impact on the adsorbents possessing a proper structure associated with an inert substrate. As in the case of the circular pellet, the concentration of COF usually remains less compared to the substrate; the textural properties of the substrate would have a great impact on both the separation and capturing processes [4]. Thus, for obtaining accurate and implementable process modelling results for scaling up the CO_2 capturing on the industrial level, proper exploration pertaining to the textural properties of different COF structures needs to be done.

3.6 Heteroatom Doping/Attachment of Functional Groups

The phenomenon of heteroatom doping is another effective way out for enhancing the CO_2 capturing capability of different porous materials, such as carbon-based adsorbents, MOFs, COFs, and other polymer materials. The incorporation of the dopant tends to change the electronic structure, surface chemistry, and pore size of the material, which in turn affects the capturing property of the material. As an example, **Figure 3.11** depicts the impact of the different dopants on the activity of the biochar material [91]. In general, mainly for the modification of the property's nitrogen, sulphur, boron, phosphorus, and fluorine, has been employed mostly. To have a proper understanding of how these dopants work, let us explore them one by one:

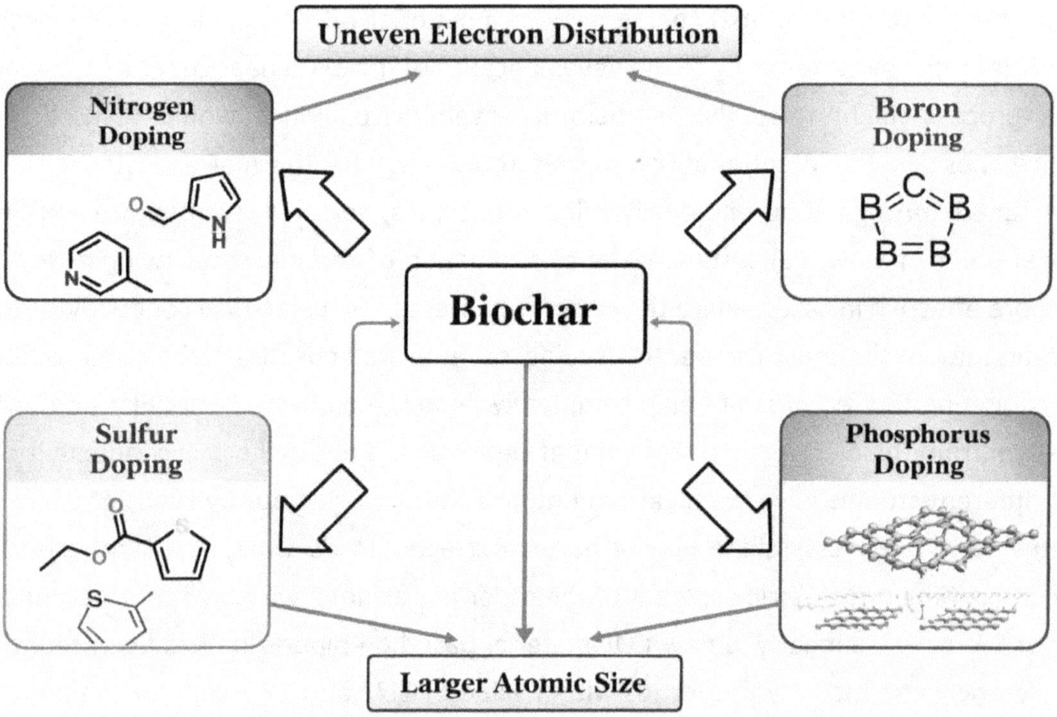

Figure 3.11: An overview of structural variations occurring due to the different heteroatom dopants. Copyright Credit: (Ref [91], Biomass Conversion and Biorefinery, Springer, Nature, 2024)

Nitrogen Doping: Among the different elements in the periodic table, nitrogen, being the next neighbour of carbon, almost possess similar atomic size as well as chemical properties to those of carbon. Because of this similarity, it tends to form strong bonds with the carbon [92]. Among the different dopants, this is one of the most studied dopants for carbon-based materials and tends to form a variety of structures by occupying different positions in the carbon structures [93]. It has been observed that the bond lengths corresponding to the nitrogen-carbon bonds are almost similar to those of the carbon–carbon bonds [94]. As a result, there is no significant difference between the spatial structure of carbon and nitrogen-doped materials. It is due to these similar characteristic features between carbon and nitrogen, due to which the incorporation of nitrogen in the carbon-based materials does not cause any significant change in the structure [95]. Owing to all the mentioned chemical similarities, nitrogen is one of the prominent elements employed for the doping and is incorporated in the carbon structure during the carbonisation [96]. It has been observed that adsorption properties of the materials get enhanced due to the doping. The doping of the nitrogen can take either of two forms in the structure, that is chemical or structure. By chemical, means the interaction with the structure through the formation of different functional groups such as amino and nitroso. These functional groups result in the enhancing the Bronsted–Lowry basicity of different carbon materials. The structural nitrogen is the occupation of the position directly in the structure, which results in transforming the structure through chemical interactions with the carbon. As a structural unit, some of the prominent structures for the doped nitrogen are graphite nitrogen, pyridine, pyrrole, etc. [97]. Nitrogen as a structural unit tends to enhance the Lewis basicity of the host materials. As a result of the enhanced basicity nature of the modified materials, the CO_2 molecules in these materials possess a strong affinity for interaction with the materials possessing small Lewis acidity, thereby resulting in enhancing the adsorption capacity of the respective carbon material [98]. Furthermore, certain carbon structures will be positively charged as a result of the electron density of nearby carbon atoms being altered by the electronegativity of nitrogen atoms [99]. In order to improve the adsorption capabilities of carbon materials, this unequal electron distribution will cause unique active sites to emerge in the carbon skeleton, altering the overall electronic band structure [100].

Boron Doping: Similar to nitrogen, the element boron resides in the periodic table before carbon, thus possesses almost similar atomic size, as well as chemical properties as that of carbon. The incorporation of the boron as a dopant does not cause any serious modification in the spatial distribution of the structure [101]. However, due to unequal distribution of electrons on the surface of the carbon materials, the incorporation of boron tends to increase the actives site, thereby resulting in enhancement of the adsorption capabilities of the desired material [102]. While comparing to nitrogen, the behaviour and thereby the mechanism will be reversed. From the electronic structure point of view, both are opposite to each other, for interaction with the carbon, boron, and even like the nitrogen interacts by posing the capability to form three chemical bonds [101]. But for the boron to form a bond, it tends to lose the electron, which is opposite to the case of nitrogen [103]. In this regard, the electronegativity of the carbon atoms adjacent to both will be altogether different from each other. A study related to observing the effect of boron on the carbon material was carried out by Li *et al.* [103]. Employing the phenolic resin and tr-isopropyl borate as the precursor, they have synthesised boron-doped carbon material through the hydrothermal carbonisation method. While analysing the material's CO_2 adsorption capabilities, it was found that the material possesses an improved adsorption capability of the order of 2.16 mmol/g, which was much better than the bare carbon material. The obtained results clearly signify that the incorporation of boron tends to enhance the CO_2 adsorption capabilities of a material.

Consequently, a tried-and-true solution with CO_2 adsorption potential is boron-doped carbon material. Although, as per the discussion, the incorporation of the boron as a dopant tends to enhance the adsorption capacity of the material, there is some lag in understanding related to the functioning mechanism of the boron. As mentioned, its incorporation tends to increase the active sites; at the same time, due to the Lewis basic nature of this structure, they interact regularly with the CO_2 molecules of weak Lewis acidic nature [104]. But due to the boron's high affinity for oxygen, these materials are prone to getting oxidised, which in turn decreases the basic nature of these materials, thereby affecting the CO_2 adsorption efficiency [101]. Thus, further research exploration is still needed to determine how to create stable boron-doped carbon materials.

Sulphur Doping: Another important dopant that is being employed for modifying the characteristic features of the CO_2 adsorbent material is the Sulphur. Due to the fact that it has a comparable charge state with that of the carbon, the interaction resulting in the formation of the chemical bond between carbon and sulphur always remains unpolarised [105]. Also, as such, there are no significant changes in the electron density of the carbon atom within the respective structure. Due to the large atomic size of the sulphur atom, the bond lengths between the sulphur and carbon are larger than that of the carbon–carbon bond [104]. As a consequence of the relatively larger bond length, there occurs a distortion, an increase in the graphite layer spacing, followed by the defects and strain in the structure [106]. The arousal of the defects along with the strain tends to change the distribution of charge, thereby resulting in the generation of more active sites, which are useful for the CO_2 adsorption. The prominent structures formed through the interaction between carbon and sulphur are the heterocyclic compounds named thiophene and thiazole [107]. Owing to the different bond lengths, the incorporation of the sulphur tends to bend the plane of configuration in the graphite, which causes the aromaticity of the respective system [106]. In nut shell, the incorporation of dopant tends to increase the active sites, which are beneficial for enhancing the efficiency of CO_2 capturing, but further exploration related to the mechanism of material transfer is required for future progress.

Phosphor Doping: Another element, phosphor, being employed for the doping resides in the nitrogen (as discussed) family in the periodic table, thereby possessing the chemical properties similar to those discussed. Due to the electronegative nature of phosphorus atoms, doping with it alters the electron density of structures, through the creation of active sites [6]. Being bigger in size than nitrogen, the respective bond length for the phosphor-carbon is also larger, which causes much more distortion, layer spacing, as well as electron density variation as compared to that of nitrogen [108]. Besides that, the actives sites created by phosphor incorporation are stretched in nature. Phosphor, as a dopant being less explored for the adsorption studies, becomes challenging to determine the responsible adsorption mechanism. In this regard, Zhou *et al.* [109] tried to examine the CO_2 capture process in different phosphide materials through the use of an electric field. They remain successful in exploring the responsible mechanism involved in different stages of the CO_2 capturing process. From the overall results, they concluded that the phosphide materials exhibit excellent characteristic features for the CO_2 capture, which even gets enhanced in the presence of an electric field. Although the improvement in the capturing efficiency has been attained through the phosphor doping, there is still a lag in the knowledge pertaining to the proper mechanism and improvement under the conventional conditions. Thus, more investigation is required for exploring the proper understanding of the mechanism, and to comprehend the impact of phosphorus's atomic charge and its spatial structure on CO_2 adsorption.

3.6.1 Photoreduction

For many years, photocatalytic reactions have garnered a lot of interest because of the constant and easily accessible solar energy source. This has increased interest in using solar energy for the degradation of organic pollutants and large-scale industrial fuel generation [110]. The whole process is named the photocatalytic process, and the spots where the reactant adsorption, disassociation, and product desorption take place in the process are known as photocatalysts. In order to perform the light energy conversion through photocatalysis, semiconductors possessing a bandgap are the most explored materials. The electrons in these materials get easily excited as soon as they receive the energy equal to or greater than the bandgap value of the material. The stimulated electron, while moving from the valence band (VB) to the conduction band (CB), leaves behind the hole vacancies called photogenerated holes.

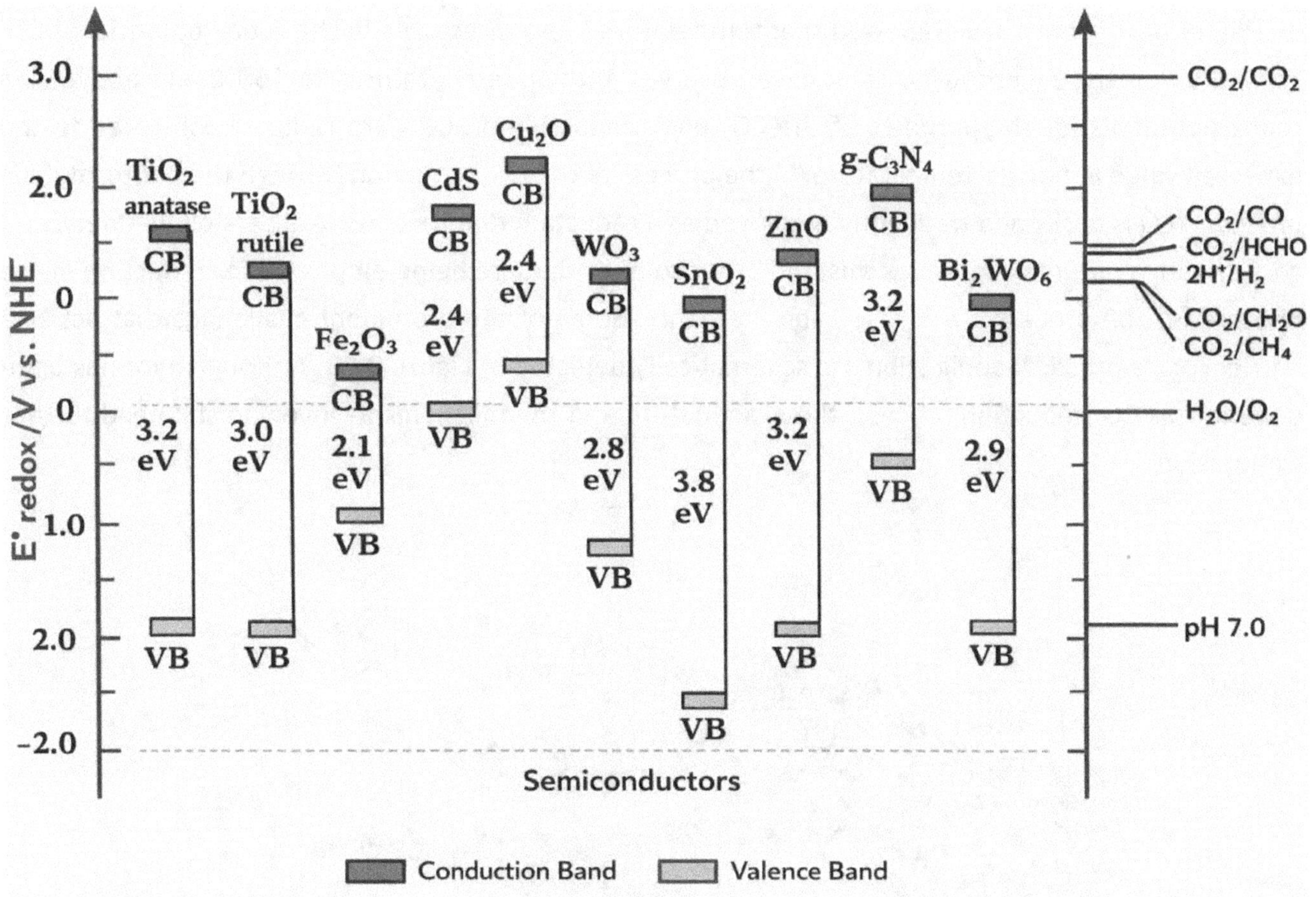

Figure 3.12: Schematic representation of the band gap for different photocatalysts at neutral pH. Copyright Credit: (Ref [112], Journal of Photochemistry and Photobiology C, Elsevier, 2015)

As soon as these semiconductor materials interact with the adsorbent materials, the generated carriers come to the surface of the semiconducting material. On the surface, these carriers trigger the initiation of the reduction and oxidation processes. At the same time, there is a possibility that these carriers may interact with their opposite counterparts present either on the surface or in the bulk, which would be detrimental for the photocatalytic process. In the process of the CO_2 reduction through the photocatalytic process, among the photogenerated carriers, the electrons take part in the reduction process of CO_2. In contrast, the holes take part in the oxidation of the reduced water [111]. It is necessary for the process to occur that the structure of the employed catalyst should be appropriate to match the reactions' redox potential. As per the principle of thermodynamics, the suitable photocatalyst for the CO_2 reduction is one whose CB potential is much greater (negative) than that of the lowest unoccupied molecular orbital (LUMO) of the redox couples involved in the process. Similarly, the photogenerated hole involved in the interaction between the semiconductor and the adsorbent will get transferred only if the energy corresponding to the highest molecular orbitals (HOMO) of the oxidised molecule is much more negative than that of the VB of the photocatalyst. The energy level, along with their equivalent redox potentials for CO_2 and the respective by-products that get generated in the CO_2 photoreduction process, are pictorially represented in **Figure 3.12** [112]. As per the study carried out by Han *et al.* [113], they have reported that temperature has a direct role in the redox potential of CO_2 as well as on the by-products. They have observed that at temperatures of 150°C and above, the redox potentials for the products of the CO_2 photoreduction process are reduced compared to the expected value at normal temperature. In the process of the CO_2 reduction through the photocatalytic process, H_2O is easily and frequently employed as a reductant due to its wide accessibility, safety, and cost-effective nature [114]. Besides this, the other sources that are being employed for reduction in the photoreduction process are H_2, CH_4, and methanol. Some of the prominent challenges that act as a barrier for its practical application are schematically depicted in **Figure 3.13**. Although a lot has been done in this CO_2 reduction process, the exact nature and the mechanisms involved still need further exploration.

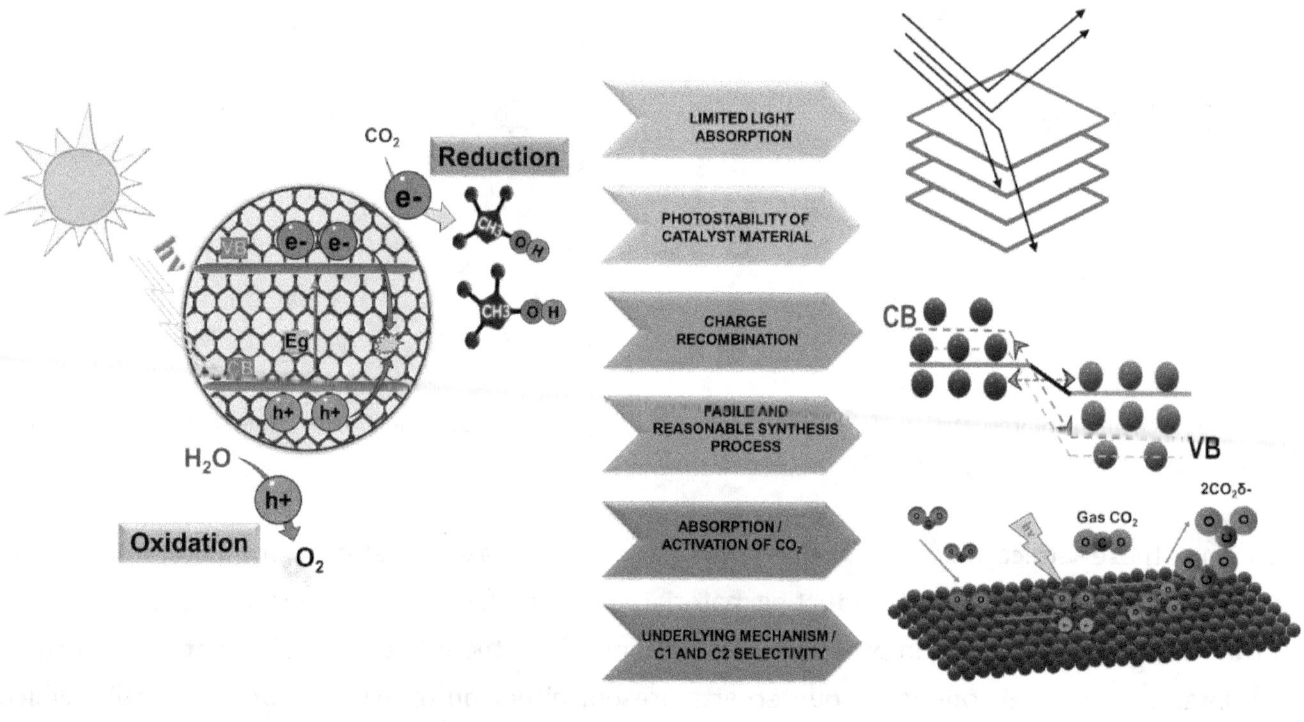

Figure 3.13: Schematic representation of the photocatalytic process and the associated challenges in CO_2 reduction, from poor light absorption to scalability issues, acting as a barrier for its practical applications. Copyright Credit: (Ref. [6], Energy & Environmental Science, RSC, 2026)

Especially related to the elementary reaction pathways, impact of the reaction conditions, selectivity, as well as efficiency of the whole process, requires a proper understanding. One of the proposed mechanisms is the one-electron reduction process. In this process, the formation of CO_2•– in the first stage leads to the progress of the subsequent reaction. With a strong negative redox potential of –1.9 V vs NHE, the CO_2/CO_2•– is thermodynamically not favourable and is thus regarded as the reaction's rate-limiting phase [115]. Another common process with many promising possibilities is multi-electron CO_2 photoreduction aided by a matching number of protons. The redox potential corresponding to the various by-products that are generated during the CO_2 photoreduction process is presented in **Table 3.3** [116]. Another challenge in the CO_2 photoreduction process is to have control over the selectivity of the products, which mainly depends on the employed photocatalyst and the reaction parameters. As an example, it is a well-established fact that N-doped TiO_2 results in the creation of CH_4 or CO in the reaction. However, in the composites of g-C_3N_4 and N-TiO_2, results in the generation of CO, and is regulated by the concentration of g-C_3N_4 in the composite [117]. On the other hand, the selectivity for the generation of CH_4 is increased in the Cu_2O and Pt co-doped TiO_2 structures [118].

Table 3.3: Products for the CO_2 reduction process with reduction potential at temperature, pressure, and pH as 25°C, 1 atm, and 7.

S.No.	Reaction	Reaction Product	Redu. Pot. (Ref. NHE)
01	$CO_2 + 2H^+ + 2e^- \rightarrow HCOOH$	Formic acid	–0.61
02	$CO_2 + 2H^+ + 2e^- \rightarrow CO + H_2O$	Carbon monoxide	–0.53
03	$CO_2 + 6H^+ + 6e^- \rightarrow CH_3OH + H_2O$	Methanol	–0.38
04	$CO_2 + 8H^+ + 8e^- \rightarrow CH_4 + 2H_2O$		–0.24
05	$2CO_2 + 12H^+ + 12e^- \rightarrow C_2H_5OH + 3H_2O$	Ethanol	–0.16
06	$CO_2 + 4H^+ + 4e^- \rightarrow HCOH + H_2O$	Formaldehyde	–0.48

Thus, through the modulation of the band structure of g-C_3N_4, it is possible to fine-tune the selectivity of the products being generated in the CO_2 photoreduction process. It is possible to fine-tune the selectivity of the CO_2 photoreduction process of g-C_3N_4 by altering the band structure. The different by-products produced in the bulk and nanosheets of the g-C_3N_4 CO_2 reduction process were mostly acetaldehyde and methane with bandgap values of 2.77 eV and 2.97 eV, respectively [119]. It has also been observed that the g-C_3N_4 materials developed through different synthesis techniques produce different by-products in the photoreduction process, thereby signifying that the synthesis method also plays a role in the process. For example, when g-C_3N_4 is developed through urea pyrolysis, the products of the photoreduction process are a mixture of CH_3OH and C_2H_5OH. At the same time, if the same materials are being synthesised through melamine pyrolysis, the product is purely C_2H_5OH. Regarding the reaction atmosphere, under the high pressure of the CO_2 liquid phase, acetic acid and alcohols remain the main reduction products [120].

3.7 Key Limitations of Surface Engineering for Carbon Capture

The phenomenon of adsorption through different materials has proven to be the best option for CO_2 capturing. So far, different adsorbent materials have been used for the same, and it has been observed that every adsorbent material has some pros and cons; as such, no perfection has been attained. Among the class of solid adsorbents, activated carbon-based solid adsorbents are the most ideal for the said adsorption process. Although these adsorbents possess many characteristic features like ease of regeneration, cost-effectiveness, and high efficiency, they still need to be made compatible with water. If this compatibility is addressed, the materials are believed to exhibit strong adsorption capacity even at low CO_2 pressure. Besides, some of the challenges pertaining to CO_2 adsorption on a solid surface that need to be addressed are as follows:

a) Most of the adsorbents have been observed to work efficiently under a laboratory-maintained atmosphere. However, in real daily life situations, such as varying temperature, moisture, presence of a mixture of gases, etc., the materials tend to lose their capability of adsorption. Thus, there is a need to prepare adsorbent materials for real-life usage.

b) The essence of being cost-effective for an adsorbent is equally important for real-life usage. Most often, the materials which aim to enhance CO_2 adsorption capacity affect the cost-effectiveness. The usage of such materials in poor nations for the purpose of achieving zero emissions will be difficult. Thus, adsorbent materials need to be cost-effective, which is challenging and needs attention.

c) All the CO_2 capture adsorbents should be assessed in terms of cost, capacity, velocity, selectivity, and regeneration. Also, only the energy-efficient adsorbents should be used to address the problem of anthropogenic CO_2 emissions.

d) Optimisation needs to be done related to the adsorbent dosage, pyrolysis temperature, metal loading, etc., by employing the right techniques.

Besides the above-mentioned aspects, which need to be taken care of, some of the other factors that can be implemented are introducing more thermodynamic analysis for efficiency evaluations, using eco-friendly materials such as agricultural wastes, biomass, and other wastes for the development of adsorbents.

3.8 Conclusion

Summarising the overall discussion, it is evident that the surface chemistry has a significant role in modifying the adsorption capabilities of the adsorbent material. Especially, the clustering behaviour of the pores has been found to play a vital role in defining the CO_2 capacity. The heteroatom doping, on the other hand, has been found to inculcate the CO_2 and acid-base interactions in the adsorbent materials. Even though the non-polar or low polarisable nature of the CO_2 tends to normalise the doping impact, the interaction has been found to contribute to enhancing the adsorption efficiency. The chemical activation, especially through KOH, works effectively in tailoring the pore structures. However, it is necessary to ensure that the activation will not lead to morphological changes and thereby not affect the pore structures. This can be attained by having meticulous control over the reaction conditions. The redox reaction has been found to improve the surface chemistry of adsorbents for CO_2 capturing through pre-oxidation as well as incorporation of some nitrogen content. The incorporation of the heteroatoms has been found to enhance the overall adsorption performance by improving the surface interactions. To regulate the efficiency of the CO_2 adsorption, it is necessary to ensure the proper doping atom, which will effectively interact with CO_2 and regulate the pore size to less than 0.8 nm. Thus, while designing the adsorbents for post-combustion CO_2 capturing, it is necessary that both the textural and chemical modifications should be taken into account. For sustainable carbon capture applications, the adsorption capacity, selectivity, and stability of carbon materials may be greatly improved by strategic surface engineering.

REFERENCES

[1] Osman, A.I., Hefny, M., Abdel Maksoud, M.I.A., Elgarahy, A.M., and Rooney, D.W. (2020) Recent advances in carbon capture storage and utilisation technologies: a review. *Environ. Chem. Lett. 2020 192*, **19** (2), 797–849.

[2] Kundu, N., and Sarkar, S. (2021) Porous organic frameworks for carbon dioxide capture and storage. *J. Environ. Chem. Eng.*, **9** (2), 105090.

[3] Fankhauser, S., Smith, S.M., Allen, M., Axelsson, K., Hale, T., Hepburn, C., Kendall, J.M., Khosla, R., Lezaun, J., Mitchell-Larson, E., Obersteiner, M., Rajamani, L., Rickaby, R., Seddon, N., and Wetzer, T. (2022) The meaning of net zero and how to get it right. *Nat. Clim. Chang.*, **12** (1), 15–21.

[4] Li, H., Dilipkumar, A., Abubakar, S., and Zhao, D. (2023) Covalent organic frameworks for CO2 capture: from laboratory curiosity to industry implementation. *Chem. Soc. Rev.*, **52** (18), 6294–6329.

[5] Ben-Mansour, R., Habib, M.A., Bamidele, O.E., Basha, M., Qasem, N.A.A., Peedikakkal, A., Laoui, T., and Ali, M. (2016) Carbon capture by physical adsorption: Materials, experimental investigations and numerical modeling and simulations - A review. *Appl. Energy*, **161**, 225–255.

[6] Madhusudhana M. Devadiga et al., Describing critical barriers in commercial deployment of CO2 capture and utilization technologies, Energy Environ. Sci., 2026,19, 743-790https://doi.org/10.1039/D6EE00134C

[7] Khurana, M., and Farooq, S. (2016) Adsorbent Screening for Postcombustion CO2 Capture: A Method Relating Equilibrium Isotherm Characteristics to an Optimum Vacuum Swing Adsorption Process Performance. *Ind. Eng. Chem. Res.*, **55** (8), 2447–2460.

[8] Zou, L., Sun, Y., Che, S., Yang, X., Wang, X., Bosch, M., Wang, Q., Li, H., Smith, M., Yuan, S., Perry, Z., and Zhou, H.C. (2017) Porous Organic Polymers for Post-Combustion Carbon Capture. *Adv. Mater.*, **29** (37), 1700229.

[9] Lin, F., Wang, Z., Zhang, Z., He, Y., Zhu, Y., Shao, J., Yuan, D., Chen, G., and Cen, K. (2020) Flue gas treatment with ozone oxidation: An overview on NOx, organic pollutants, and mercury. *Chem. Eng. J.*, **382**, 123030.

[10] Cormos, A.M., Dragan, S., Petrescu, L., Sandu, V., and Cormos, C.C. (2020) Techno-economic and environmental evaluations of decarbonized fossil-intensive industrial processes by reactive absorption and adsorption CO2 capture systems. *Energies*, **13** (5), 1268.

[11] House, K.Z., Baclig, A.C., Ranjan, M., Van Nierop, E.A., Wilcox, J., and Herzog, H.J. (2011) Economic and energetic analysis of capturing CO2from ambient air. *Proc. Natl. Acad. Sci. U. S. A.*, **108** (51), 20428–20433.

[12] Haertel, C.J.J., McNutt, M., Ozkan, M., Aradóttir, E.S.P., Valsaraj, K.T., Sanberg, P.R., Talati, S., and Wilcox, J. (2021) The promise of scalable direct air capture. *Chem*, **7** (11), 2831–2834.

[13] Davis, M.E. (2002) Ordered porous materials for emerging applications. *Nature*, **417** (6891), 813–821.

[14] Slater, A.G., and Cooper, A.I. (2015) Function-led design of new porous materials. *Science (80).*, **348** (6238), aaa8075.

[15] Zhang, X., Yan, M., Chen, P., Li, J., Li, Y., Li, H., Liu, X., Chen, Z., Yang, H., Wang, S., Wang, J., Tang, Z., Huang, Q., Lei, J., Hayat, T., Liu, Z., Mao, L., Duan, T., and Wang, X. (2025) Emerging MOFs, COFs, and their derivatives for energy and environmental applications. *Innov.*, **6** (2), 100778.

[16] Aksu, G.O., Erucar, I., Haslak, Z.P., and Keskin, S. (2022) Accelerating discovery of COFs for CO2 capture and H2 purification using structurally guided computational screening. *Chem. Eng. J.*, **427**, 131574.

[17] Rajagopalan, A.K., Avila, A.M., and Rajendran, A. (2016) Do adsorbent screening metrics predict process performance? A process optimisation based study for post-combustion capture of CO_2. *Int. J. Greenh. Gas Control*, **46**, 76–85.

[18] Patra, K., and Pal, H. (2024) Recent advances in porous organic polymers (POPs): The emerging sorbent materials with promises towards toxic and radionuclides metal ions separations. *Mater. Today Sustain.*, **27**, 100799.

[19] Dziejarski, B., Krzyżyńska, R., and Andersson, K. (2023) Current status of carbon capture, utilization, and storage technologies in the global economy: A survey of technical assessment. *Fuel*, **342**, 127776.

[20] Karimi, M., Jodaei, A., Khajvandi, A., Sadeghinik, A., and Jahandideh, R. (2018) In-situ capture and conversion of atmospheric CO2 into nano-CaCO3 using a novel pathway based on deep eutectic choline chloride-calcium chloride. *J. Environ. Manage.*, **206**, 516–522.

[21] Meyer, D. (2016) CARBON CAPTURE. *Sci. Am.*, **314** (5), 6.

[22] He, Y., Sun, J., Yao, W., Lu, K., Liu, D., Xie, H., Huang, C., and Jia, N. (2023) A self-powered photoelectrochemical molecular imprinted sensor for chloroquine phosphate with enhanced cathodic photocurrent via stepped energy band alignment engineering. *Chem. Eng. J.*, **451**, 138748.

[23] Zhang, Z., Cano, Z.P., Luo, D., Dou, H., Yu, A., and Chen, Z. (2019) Rational design of tailored porous carbon-based materials for CO2 capture. *J. Mater. Chem. A*, **7** (37), 20985–21003.

[24] Bernardo, P., Drioli, E., and Golemme, G. (2009) Membrane gas separation: A review/state of the art. *Ind. Eng. Chem. Res.*, **48** (10), 4638–4663.

[25] Yong, J., Chen, F., Yang, Q., Huo, J., and Hou, X. (2017) Superoleophobic surfaces. *Chem. Soc. Rev.*, **46** (14), 4168–4217.

[26] Seifritz, W. (1990) CO_2 disposal by means of silicates [12]. *Nature*, **345** (6275), 486.

[27] Olajire, A.A. (2013) A review of mineral carbonation technology in sequestration of CO_2. *J. Pet. Sci. Eng.*, **109**, 364–392.

[28] Fagerlund, J., Teir, S., Nduagu, E., and Zevenhoven, R. (2009) Carbonation of magnesium silicate mineral using a pressurised gas/solid process. *Energy Procedia*, **1** (1), 4907–4914.

[29] Fagerlund, J., Nduagu, E., Romão, I., and Zevenhoven, R. (2010) A stepwise process for carbon dioxide sequestration using magnesium silicates. *Front. Chem. Eng. China*, **4** (2), 133–141.

[30] Herce, C., Stendardo, S., and Cortés, C. (2015) Increasing CO_2 carrying capacity of dolomite by means of thermal stabilization by triggered calcination. *Chem. Eng. J.*, **262**, 18–28.

[31] Nduagu, E., Björklöf, T., Fagerlund, J., Wärn, J., Geerlings, H., and Zevenhoven, R. (2012) Production of magnesium hydroxide from magnesium silicate for the purpose of CO_2 mineralisation - Part 1: Application to Finnish serpentinite. *Miner. Eng.*, **30**, 75–86.

[32] Assima, G.P., Larachi, F., Molson, J., and Beaudoin, G. (2014) Impact of temperature and oxygen availability on the dynamics of ambient CO2 mineral sequestration by nickel mining residues. *Chem. Eng. J.*, **240**, 394–403.

[33] Azdarpour, A., Asadullah, M., Junin, R., Manan, M., Hamidi, H., and Mohammadian, E. (2014) Direct carbonation of red gypsum to produce solid carbonates. *Fuel Process. Technol.*, **126**, 429–434.

[34] Sanna, A., Uibu, M., Caramanna, G., Kuusik, R., and Maroto-Valer, M.M. (2014) A review of mineral carbonation technologies to sequester CO2. *Chem. Soc. Rev.*, **43** (23), 8049–8080.

[35] Azdarpour, A., Junin, R., Asadullah, M., Hamidi, H., Manan, M., and Mohamad Daud, A.R. (2014) Calcium Carbonate Production through Direct Mineral Carbon Dioxide Sequestration. *Appl. Mech. Mater.*, **699**, 1020–1025.

[36] Schmidt, R.B., and Rmanosky, B. (2001) Program Overviews: CO2 Mineral Sequestration. *MC Work. (Ed.), Pittsburgh, USA*.

[37] Azdarpour, A., Rahmani, M., Asadullah, R., Junin, M., Manan, R., and Mohsin (2013) Carbon dioxide mineral carbonation using natural minerals as feedstocks. *Int. Conf. Front. Environ. Energy Biosci.*, (December 2014), 318–323.

[38] Bobicki, E.R., Liu, Q., Xu, Z., and Zeng, H. (2012) Carbon capture and storage using alkaline industrial wastes. *Prog. Energy Combust. Sci.*, **38** (2), 302–320.

[39] Soo, C.L., Ho, J.C., Soo, J.L., Bo, Y.C., Chang, K.Y., Joong, B.L., Chong, K.R., and Jae, C.K. (2008) Development of regenerate MgO-based sorbent promoted with K_2CO_3 for CO_2 capture at low temperatures. *Environ. Sci. Technol.*, **42** (8), 2736–2741.

[40] Lim, M., Han, G.C., Ahn, J.W., and You, K.S. (2010) Environmental remediation and conversion of carbon dioxide (CO_2) into useful green products by accelerated carbonation technology. *Int. J. Environ. Res. Public Health*, **7** (1), 203–228.

[41] Alexander, G., Mercedes Maroto-Valer, M., and Gafarova-Aksoy, P. (2007) Evaluation of reaction variables in the dissolution of serpentine for mineral carbonation. *Fuel*, **86** (1–2), 273–281.

[42] Zhao, L., Sang, L., Jun, C., Ji, J., and Teng, H.H. (2010) Aqueous carbonation of natural brucite: Relevance to CO2 sequestration. *Environ. Sci. Technol.*, **44** (1), 406–411.

[43] Zevenhoven, R., Teir, S., and Eloneva, S. (2008) Heat optimisation of a staged gas-solid mineral carbonation process for long-term CO_2 storage. *Energy*, **33** (2), 362–370.

[44] Calabrò, A., Deiana, P., Fiorini, P., Girardi, G., and Stendardo, S. (2008) Possible optimal configurations for the ZECOMIX high efficiency zero emission hydrogen and power plant. *Energy*, **33** (6), 952–962.

[45] Park, A.H.A., Jadhav, R., and Fan, L.S. (2003) CO_2 mineral sequestration: Chemically enhanced aqueous carbonation of serpentine. *Can. J. Chem. Eng.*, **81** (3–4), 885–890.

[46] GERDEMANN, S., DAHLIN, D., and OCONNOR, W. (2003) *Carbon Dioxide Sequestration by Aqueous Mineral Carbonation of Magnesium Silicate Minerals*, Elsevier.

[47] Wu, J., Zhu, X., Yang, F., Wang, R., and Ge, T. (2022) Shaping techniques of adsorbents and their applications in gas separation: a review. *J. Mater. Chem. A*, **10** (43), 22853–22895.

[48] Quang, D.V., Dindi, A., Rayer, A. V., Hadri, N. El, Abdulkadir, A., and Abu-Zahra, M.R.M. (2015) Effect of moisture on the heat capacity and the regeneration heat required for CO_2 capture process using PEI impregnated mesoporous precipitated silica. *Greenh. Gases Sci. Technol.*, **5** (1), 91–101.

[49] Yong, Z., Mata, V., and Rodrigues, A.E. (2002) Adsorption of carbon dioxide at high temperature - A review. *Sep. Purif. Technol.*, **26** (2–3), 195–205.

[50] Sevilla, M., Falco, C., Titirici, M.M., and Fuertes, A.B. (2012) High-performance CO_2 sorbents from algae. *RSC Adv.*, **2** (33), 12792–12797.

[51] Choma, J., Osuchowski, L., Marszewski, M., Dziura, A., and Jaroniec, M. (2016) Developing microporosity in Kevlar®-derived carbon fibers by CO_2 activation for CO_2 adsorption. *J. CO_2 Util.*, **16**, 17–22.

[52] Lee, M.S., and Park, S.J. (2015) Silica-coated multi-walled carbon nanotubes impregnated with polyethyleneimine for carbon dioxide capture under the flue gas condition. *J. Solid State Chem.*, **226**, 17–23.

[53] Chiang, Y.C., and Juang, R.S. (2017) Surface modifications of carbonaceous materials for carbon dioxide adsorption: A review. *J. Taiwan Inst. Chem. Eng.*, **71**, 214–234.

[54] Presser, V., McDonough, J., Yeon, S.H., and Gogotsi, Y. (2011) Effect of pore size on carbon dioxide sorption by carbide derived carbon. *Energy Environ. Sci.*, **4** (8), 3059–3066.

[55] Hao, W., Björkman, E., Lilliestråle, M., and Hedin, N. (2013) Activated carbons prepared from hydrothermally carbonized waste biomass used as adsorbents for CO2. *Appl. Energy*, **112**, 526–532.

[56] Wickramaratne, N.P., and Jaroniec, M. (2013) Importance of small micropores in CO_2 capture by phenolic resin-based activated carbon spheres. *J. Mater. Chem. A*, **1** (1), 112–116.

[57] Wang, Z., Zhan, L., Ge, M., Xie, F., Wang, Y., Qiao, W., Liang, X., and Ling, L. (2011) Pith based spherical activated carbon for CO2 removal from flue gases. *Chem. Eng. Sci.*, **66** (22), 5504–5511.

[58] Chowdhury, S., and Balasubramanian, R. (2016) Highly efficient, rapid and selective CO_2 capture by thermally treated graphene nanosheets. *J. CO_2 Util.*, **13**, 50–60.

[59] Yu, L., Falco, C., Weber, J., White, R.J., Howe, J.Y., and Titirici, M.M. (2012) Carbohydrate-derived hydrothermal carbons: A thorough characterization study. *Langmuir*, **28** (33), 12373–12383.

[60] Singh, V.K., and Anil Kumar, E. (2016) Measurement and analysis of adsorption isotherms of CO2 on activated carbon. *Appl. Therm. Eng.*, **97**, 77–86.

[61] Tan, Y.L., Islam, M.A., Asif, M., and Hameed, B.H. (2014) Adsorption of carbon dioxide by sodium hydroxide-modified granular coconut shell activated carbon in a fixed bed. *Energy*, **77**, 926–931.

[62] Arami-Niya, A., Rufford, T.E., and Zhu, Z. (2016) Activated carbon monoliths with hierarchical pore structure from tar pitch and coal powder for the adsorption of CO_2, CH_4 and N_2. *Carbon N. Y.*, **103**, 115–124.

[63] Alabadi, A., Razzaque, S., Yang, Y., Chen, S., and Tan, B. (2015) Highly porous activated carbon materials from carbonized biomass with high CO_2 capturing capacity. *Chem. Eng. J.*, **281**, 606–612.

[64] Li, D., Ma, T., Zhang, R., Tian, Y., and Qiao, Y. (2015) Preparation of porous carbons with high low-pressure CO_2 uptake by KOH activation of rice husk char. *Fuel*, **139**, 68–70.

[65] Meng, L.Y., and Park, S.J. (2010) Effect of heat treatment on CO_2 adsorption of KOH-activated graphite nanofibers. *J. Colloid Interface Sci.*, **352** (2), 498–503.

[66] Lee, S.Y., and Park, S.J. (2013) Determination of the optimal pore size for improved CO_2 adsorption in activated carbon fibers. *J. Colloid Interface Sci.*, **389** (1), 230–235.

[67] Gopalan, J., Buthiyappan, A., and Abdul Raman, A.A. (2022) Insight into metal-impregnated biomass based activated carbon for enhanced carbon dioxide adsorption: A review. *J. Ind. Eng. Chem.*, **113**, 72–95.

[68] Serafin, J., Ouzzine, M., Cruz Junior, O.F., and Sreńscek-Nazzal, J. (2021) Preparation of low-cost activated carbons from amazonian nutshells for CO2 storage. *Biomass and Bioenergy*, **144**, 105925.

[69] Yang, F., Zhang, S., Sun, Y., Cheng, K., Li, J., and Tsang, D.C.W. (2018) Fabrication and characterization of hydrophilic corn stalk biochar-supported nanoscale zero-valent iron composites for efficient metal removal. *Bioresour. Technol.*, **265**, 490–497.

[70] Hosseini, S., Bayesti, I., Marahel, E., Eghbali Babadi, F., Chuah Abdullah, L., and Choong, T.S.Y. (2015) Adsorption of carbon dioxide using activated carbon impregnated with Cu promoted by zinc. *J. Taiwan Inst. Chem. Eng.*, **52**, 109–117.

[71] Cha, J.S., Choi, J.C., Ko, J.H., Park, Y.K., Park, S.H., Jeong, K.E., Kim, S.S., and Jeon, J.K. (2010) The low-temperature SCR of NO over rice straw and sewage sludge derived char. *Chem. Eng. J.*, **156** (2), 321–327.

[72] Wang, Z., Shi, M., Li, J., and Zheng, Z. (2014) Influence of moderate pre-oxidation treatment on the physical, chemical and phosphate adsorption properties of iron-containing activated carbon. *J. Environ. Sci. (China)*, **26** (3), 519–528.

[73] Qamar, S., Lei, F., Liang, L., Gao, S., Liu, K., Sun, Y., Ni, W., and Xie, Y. (2016) Ultrathin TiO_2 flakes optimizing solar light driven CO_2 reduction. *Nano Energy*, **26**, 692–698.

[74] Asadi-Sangachini, Z., Galangash, M.M., Younesi, H., and Nowrouzi, M. (2019) The feasibility of cost-effective manufacturing activated carbon derived from walnut shells for large-scale CO_2 capture. *Environ. Sci. Pollut. Res.*, **26** (26), 26542–26552.

[75] Jeguirim, M., Belhachemi, M., Limousy, L., and Bennici, S. (2018) Adsorption/reduction of nitrogen dioxide on activated carbons: Textural properties versus surface chemistry – A review. *Chem. Eng. J.*, **347**, 493–504.

[76] Yargicoglu, E.N., Sadasivam, B.Y., Reddy, K.R., and Spokas, K. (2015) Physical and chemical characterization of waste wood derived biochars. *Waste Manag.*, **36**, 256–268.

[77] Wang, F., Jin, L., Guo, C., Min, L., Zhang, P., Sun, H., Zhu, H., and Zhang, C. (2021) Enhanced heavy metals sorption by modified biochars derived from pig manure. *Sci. Total Environ.*, **786**, 147595.

[78] Danaci, D., Bui, M., Mac Dowell, N., and Petit, C. (2020) Exploring the limits of adsorption-based CO_2 capture using MOFs with PVSA-from molecular design to process economics. *Mol. Syst. Des. Eng.*, **5** (1), 212–231.

[79] Farmahini, A.H., Friedrich, D., Brandani, S., and Sarkisov, L. (2020) Exploring new sources of efficiency in process-driven materials screening for post-combustion carbon capture. *Energy Environ. Sci.*, **13** (3), 1018–1037.

[80] Gui, B., Liu, X., Cheng, Y., Zhang, Y., Chen, P., He, M., Sun, J., and Wang, C. (2022) Tailoring the Pore Surface of 3D Covalent Organic Frameworks via Post-Synthetic Click Chemistry. *Angew. Chemie*, **134** (2), e202113852.

[81] Zhao, S., Dong, B., Ge, R., Wang, C., Song, X., Ma, W., Wang, Y., Hao, C., Guo, X., and Gao, Y. (2016) Channel-wall functionalization in covalent organic frameworks for the enhancement of CO_2 uptake and CO_2/N_2 selectivity. *RSC Adv.*, **6** (45), 38774–38781.

[82] Moosavi, S.M., Novotny, B.Á., Ongari, D., Moubarak, E., Asgari, M., Kadioglu, Ö., Charalambous, C., Ortega-Guerrero, A., Farmahini, A.H., Sarkisov, L., Garcia, S., Noé, F., and Smit, B. (2022) A data-science approach to predict the heat capacity of nanoporous materials. *Nat. Mater.*, **21** (12), 1419–1425.

[83] Singh, G., Lee, J., Karakoti, A., Bahadur, R., Yi, J., Zhao, D., Albahily, K., and Vinu, A. (2020) Emerging trends in porous materials for CO2capture and conversion. *Chem. Soc. Rev.*, **49** (13), 4360–4404.

[84] Huang, N., Chen, X., Krishna, R., and Jiang, D. (2015) Two-dimensional covalent organic frameworks for carbon dioxide capture through channel-wall functionalization. *Angew. Chemie - Int. Ed.*, **54** (10), 2986–2990.

[85] Freitas, S.K.S., Borges, R.S., Merlini, C., Barra, G.M.O., and Esteves, P.M. (2017) Thermal Conductivity of Covalent Organic Frameworks as a Function of Their Pore Size. *J. Phys. Chem. C*, **121** (48), 27247–27252.

[86] Ma, H., Aamer, Z., and Tian, Z. (2021) Ultrahigh thermal conductivity in three-dimensional covalent organic frameworks. *Mater. Today Phys.*, **21**, 100536.

[87] Giri, A., and Hopkins, P.E. (2021) Heat Transfer Mechanisms and Tunable Thermal Conductivity Anisotropy in Two-Dimensional Covalent Organic Frameworks with Adsorbed Gases. *Nano Lett.*, **21** (14), 6188–6193.

[88] Keller, L., Lohaus, T., Abduly, L., Hadler, G., and Wessling, M. (2019) Electrical swing adsorption on functionalized hollow fibers. *Chem. Eng. J.*, **371**, 107–117.

[89] Evans, A.M., Giri, A., Sangwan, V.K., Xun, S., Bartnof, M., Torres-Castanedo, C.G., Balch, H.B., Rahn, M.S., Bradshaw, N.P., Vitaku, E., Burke, D.W., Li, H., Bedzyk, M.J., Wang, F., Brédas, J.L., Malen, J.A., McGaughey, A.J.H., Hersam, M.C., Dichtel, W.R., and Hopkins, P.E. (2021) Thermally conductive ultra-low-k dielectric layers based on two-dimensional covalent organic frameworks. *Nat. Mater.*, **20** (8), 1142–1148.

[90] Yue, Y., Li, H., Chen, H., and Huang, N. (2022) Piperazine-Linked Covalent Organic Frameworks with High Electrical Conductivity. *J. Am. Chem. Soc.*, **144** (7), 2873–2878.

[91] Sun, Y., Jia, J., Huo, L., Zhao, L., Yao, Z., and Liu, Z. (2024) Heteroatom-doped biochar for CO2 adsorption: a review of heteroatoms, doping methods, and functions. *Biomass Convers. Biorefinery*, **14** (14), 15237–15249.

[92] Li, L., Wang, X.F., Zhong, J.J., Qian, X., Song, S.L., Zhang, Y.G., and Li, D.H. (2018) Nitrogen-enriched porous polyacrylonitrile-based carbon fibers for CO_2 Capture. *Ind. Eng. Chem. Res.*, **57** (34), 11608–11616.

[93] Xiao, J., Yuan, X., Zhang, T.C., Ouyang, L., and Yuan, S. (2022) Nitrogen-doped porous carbon for excellent CO_2 capture: A novel method for preparation and performance evaluation. *Sep. Purif. Technol.*, **298**, 121602.

[94] Jin, J., Gu, Y.J., Man, C.W.Y., Cheng, J., Xu, Z., Zhang, Y., Wang, H., Lee, V.H.Y., Cheng, S.H., and Wong, W.T. (2011) Polymer-coated $NaYF_4$:Yb^{3+}, Er^{3+} upconversion nanoparticles for charge-dependent cellular imaging. *ACS Nano*, **5** (10), 7838–7847.

[95] Cao, Z., Liu, S., Xu, K., Mao, Y., Wu, Y., and Mao, Q. (2021) Predictable interfacial mass transfer intensification of Sn-N doped multichannel hollow carbon nanofibers for the CO2electro-reduction reaction. *Sustain. Energy Fuels*, **5** (12), 3097–3101.

[96] Ning, H., Yang, Z., Wang, D., Meng, Z., Li, Y., Ju, X., and Wang, C. (2021) Graphene-based semi-coke porous carbon with N-rich hierarchical sandwich-like structure for efficient separation of CO_2/N_2. *Microporous Mesoporous Mater.*, **311**, 110700.

[97] Mukhtar, A., Saqib, S., Mellon, N.B., Rafiq, S., Babar, M., Ullah, S., Muhammad, N., Khan, A.L., Ayoub, M., Ibrahim, M., Maqsood, K., Bustam, M.A., Al-Sehemi, A.G., Klemeš, J.J., Asif, S., and Bokhari, A. (2020) A review on CO_2 capture via nitrogen-doped porous polymers and catalytic conversion as a feedstock for fuels. *J. Clean. Prod.*, **277**, 123999.

[98] Li, Q., Lu, T., Wang, L., Pang, R., Shao, J., Liu, L., and Hu, X. (2021) Biomass based N-doped porous carbons as efficient CO_2 adsorbents and high-performance supercapacitor electrodes. *Sep. Purif. Technol.*, **275**, 119204.

[99] Leng, L., Xu, S., Liu, R., Yu, T., Zhuo, X., Leng, S., Xiong, Q., and Huang, H. (2020) Nitrogen containing functional groups of biochar: An overview. *Bioresour. Technol.*, **298**, 122286.

[100] Li, W., Cui, H., Xu, J., and Shi, J. (2022) N-doped porous carbon prepared from filter paper for CO2 capture. *Chem. Phys. Lett.*, **787**, 139275.

[101] Shen, Z., Song, Y., Yin, C., Luo, X., Wang, Y., and Li, X. (2021) Construction of hierarchically porous 3D graphene-like carbon material by B, N co-doping for enhanced CO_2 capture. *Microporous Mesoporous Mater.*, **322**, 111158.

[102] Ibarra-Rodríguez, M., and Sánchez, M. (2020) Graphitic carbon nitride functionalized with four boron atoms for adsorption and separation of CO_2/CH_4: DFT calculations. *Adsorption*, **26** (4), 597–605.

[103] Li, J., Shi, C., and Bao, A. (2021) Design of boron-doped mesoporous carbon materials for multifunctional applications: Dye adsorption and CO_2 capture. *J. Environ. Chem. Eng.*, **9** (3), 105250.

[104] Wang, H., Shao, Y., Mei, S., Lu, Y., Zhang, M., Sun, J.K., Matyjaszewski, K., Antonietti, M., and Yuan, J. (2020) Polymer-Derived Heteroatom-Doped Porous Carbon Materials. *Chem. Rev.*, **120** (17), 9363–9419.

[105] Liu, S., Cai, Y., Zhao, X., Liang, Y., Zheng, M., Hu, H., Dong, H., Jiang, S., Liu, Y., and Xiao, Y. (2017) Sulfur-doped nanoporous carbon spheres with ultrahigh specific surface area and high electrochemical activity for supercapacitor. *J. Power Sources*, **360**, 373–382.

[106] Su, W., Yao, L., Ran, M., Sun, Y., Liu, J., and Wang, X. (2018) Adsorption Properties of N_2, CH_4, and CO_2 on Sulfur-Doped Microporous Carbons. *J. Chem. Eng. Data*, **63** (8), 2914–2920.

[107] Wohlgemuth, S.A., White, R.J., Willinger, M.G., Titirici, M.M., and Antonietti, M. (2012) A one-pot hydrothermal synthesis of sulfur and nitrogen doped carbon aerogels with enhanced electrocatalytic activity in the oxygen reduction reaction. *Green Chem.*, **14** (5), 1515–1523.

[108] Khan, A.A., Ahmad, I., and Ahmad, R. (2020) Influence of electric field on CO_2 removal by P-doped C60-fullerene: A DFT study. *Chem. Phys. Lett.*, **742**, 137155.

[109] Zhou, S., Wang, M., Wang, J., Xin, H., Liu, S., Wang, Z., Wei, S., and Lu, X. (2020) Carbon phosphides: promising electric field controllable nanoporous materials for CO2capture and separation. *J. Mater. Chem. A*, **8** (19), 9970–9980.

[110] Chandrasekaran, S., Chung, J.S., Kim, E.J., and Hur, S.H. (2016) Advanced nano-structured materials for photocatalytic water splitting. *J. Electrochem. Sci. Technol.*, **7** (1), 1–12.

[111] Yuan, L., and Xu, Y.J. (2015) Photocatalytic conversion of CO_2 into value-added and renewable fuels. *Appl. Surf. Sci.*, **342**, 154–167.

[112] Ola, O., and Maroto-Valer, M.M. (2015) Review of material design and reactor engineering on TiO_2 photocatalysis for CO_2 reduction. *J. Photochem. Photobiol. C Photochem. Rev.*, **24**, 16–42.

[113] Han, B., Wei, W., Chang, L., Cheng, P., and Hu, Y.H. (2016) Efficient Visible Light Photocatalytic CO_2 Reforming of CH_4. *ACS Catal.*, **6** (2), 494–497.

[114] Wang, Z., Teramura, K., Huang, Z., Hosokawa, S., Sakata, Y., and Tanaka, T. (2016) Tuning the selectivity toward CO evolution in the photocatalytic conversion of CO_2 with H_2O through the modification of Ag-loaded Ga_2O_3 with a $ZnGa_2O_4$ layer. *Catal. Sci. Technol.*, **6** (4), 1025–1032.

[115] Tu, W., Zhou, Y., and Zou, Z. (2014) Photocatalytic conversion of CO_2 into renewable hydrocarbon fuels: State-of-the-art accomplishment, challenges, and prospects. *Adv. Mater.*, **26** (27), 4607–4626.

[116] Wang, W.-N., Soulis, J., Yang, Y.J., and Biswas, P. (2014) Comparison of CO_2 Photoreduction Systems: A Review. *Aerosol Air Qual. Res.*, **14** (2), 533–549.

[117] Zhou, S., Liu, Y., Li, J., Wang, Y., Jiang, G., Zhao, Z., Wang, D., Duan, A., Liu, J., and Wei, Y. (2014) Facile in situ synthesis of graphitic carbon nitride (g-C_3N_4)-N-TiO_2 heterojunction as an efficient photocatalyst for the selective photoreduction of CO_2 to CO. *Appl. Catal. B Environ.*, **158–159**, 20–29.

[118] Xiong, Z., Lei, Z., Kuang, C.C., Chen, X., Gong, B., Zhao, Y., Zhang, J., Zheng, C., and Wu, J.C.S. (2017) Selective photocatalytic reduction of CO_2 into CH_4 over Pt-Cu_2O TiO_2 nanocrystals: The interaction between Pt and Cu2O cocatalysts. *Appl. Catal. B Environ.*, **202**, 695–703.

[119] Niu, P., Yang, Y., Yu, J.C., Liu, G., and Cheng, H.M. (2014) Switching the selectivity of the photoreduction reaction of carbon dioxide by controlling the band structure of a g-C_3N_4 photocatalyst. *Chem. Commun.*, **50** (74), 10837–10840.

[120] Mizuno, T., Adachi, K., Ohta, K., and Saji, A. (1996) Effect of CO_2 pressure on photocatalytic reduction of CO_2 using TiO_2 in aqueous solutions. *J. Photochem. Photobiol. A Chem.*, **98** (1–2), 87–90.

Chapter 4

NANOCOMPOSITES FOR CARBON CAPTURE TECHNOLOGIES

Abstract: The continuous rise of the CO_2 concentration in the atmosphere results in the evident escalation of environmental challenges. These challenges demand that the scientific community develop different innovations as well as mitigation strategies. Among the different efforts, carbon capture, its utility, and storage are the advanced efforts in this direction. These processes employ nanocomposite materials, which have proven to be a transformative class of materials. These materials exhibit efficient properties, thereby effectively replacing the traditional conventional approaches. This chapter provides an overview of the role played by nanocomposite materials in the advancement of growing carbon-capturing technology. The entitled chapter firstly provides a brief overview of nanocomposites and their properties, owing to which they offer excellent results in CO_2 capturing. The section is followed by an overview of the classification of nanocomposites and their description. The synthesis procedure through which these different classes of nanocomposites are being generated is discussed in detail. Afterwards, the chapter provides a discussion about the thermo-mechanical and morphological properties of these nanocomposites and their correlation through which their properties can be modified. Besides discussing the different mechanisms for capturing, a critical overview of the challenges faced in the different stages from synthesis to the industrial scale is presented as well. This comprehensive report aims to act as a solid foundation source for providing the basic understanding of the current state, progress, future, and challenges in the nanocomposites for CO_2 capturing technology.

4.1 Introduction

The ever-rising amount of CO_2 in the atmosphere is due to anthropogenic activities, which include industrialisation, burning of fossil fuels, and widespread deforestation [1]. The rising concentration of CO_2 in the atmosphere is among the most concerning issues of the current century, as it poses a threat to all kinds of life on the planet. The immediate cause of its rising content in the atmosphere is global warming, which causes severe climate changes, thus forcing the world research community for designing the innovation and effective technologies to mitigate emissions [2]. Among the different attempts and efforts made so far on different platforms, CCUS are considered to be a viable option for reducing the CO_2 in the atmosphere. This technology has been recognised globally by the research community, and different progress is being made in this domain [3]. The mentioned technologies prevent the emission of CO_2 in the atmosphere by directly capturing the CO_2 from the emission sources, such as power plants, industries or from the ambient air, followed by its storage and conversion into useful energy. This employed CCUS technology is considered a favourable option for attaining the aim of zero-carbon emission apart from the other approaches. Besides this, it is considered favourable for the low-carbon global economy [4].

For the capturing technology, different materials are being employed; nanocomposites are considered to be the revolutionary class of materials in this domain. These materials exhibit unique properties owing to their engineering, through which these materials are being developed using a synergetic combination of different nanoscale components. These materials have changed the scenario of the CO_2 capturing technology as they have proved to be smart materials which make the process effective, selective, and affordable. These materials have totally replaced the conventional approaches of the CO_2 capturing technology, which faces multiple challenges [4]. As an example among the first-generation technologies, those based on the amine scrubbing were not energy-efficient owing to the huge energy requirements in the solvent regeneration. This, in turn, induces huge energy requirements and thereby lowers the efficiency by decreasing the net output power. Besides that, the systems developed on the basis of chemical solvents face degradation with time, which often results in the formation of by-products that are not suitable for the environment. The other traditional approaches face challenges such as low adsorption capacity, weak selectivity in an atmosphere with moisture, gases like nitrogen, etc., as well as high operational cost. These limitations have encouraged researchers across the globe for the development of new smart materials. These materials are aimed to possess high adsorption capacities, strong selectivity, eco-friendly, and cost-effective [4, 5]. Looking for the materials with all these properties has led to the development of a new interdisciplinary field of science called nanotechnology, which involves the tuning of the materials' properties at the atomic and molecular levels. This field have results in the development of efficient materials for versatile phenomena through engineering at the nanoscale regime, thereby attaining the name of "nano-green revolution." The capability of engineering the material at the nanoscale level has made it possible to develop materials with high surface area to volume ratios, tuneable pore size, and tuning of the surface chemistry properties. Materials with all these properties have been found to prove excellent for the CCUS [6].

The class of materials called nanocomposites have emerged as an intersection of materials science and nanotechnology. These are the materials with multiphases, with each phase having its own dimensionality of size less than 100 nms or some of these are composed of a repeated network of different phases [1, 4]. In view of the carbon-capturing technology, different nanocomposite materials have been developed so far, such as carbon nanotubes, MOFs, ceramic polymers, etc. [3]. The aim of designing the nanocomposite materials is to develop a material with coactions of features exhibited in individual components. As an example, the carbon nanotubes (CNTs), which are known for their mechanical strength, can be combined with MOFs, which possess high porosity and selectivity, resulting in the development of a hybrid material which will be efficient for CO_2 capturing technology. This approach of developing the different composites is excellent for both post-and pre-combustion processes. The development of these materials has altogether changed the paradigm of carbon-capturing technology, by shifting from the generic version to one-size-fits-all approaches for designing the materials with efficient properties and performances [3].

There is a diversity of materials that are being engineered at the nanoscale regime for designing different nanocomposites for the carbon-capturing technology. Among the different nanocomposites, the carbon-based nanocomposite materials employ the different carbon allotropes for designing the different nanostructures [7]. Among this class, the materials like graphene (sheet-like structure) and carbon nanotubes (rolled sheets of graphene) possess excellent properties such as high surface area, thermal and chemical stability, etc., suitable for the materials to act as a nanofiller. On being incorporated with the polymer matrix, these result in the generation of tortuous structures within the matrix that help in the separation of gases, and even they themselves act as adsorbent surfaces. Similarly, other carbon-based materials like activated carbon and biochar, with biomass as the source has proven to be cost-effective nanocomposite materials with affinity for CO_2 [8]. Another well-known class of nanocomposite materials are MOFs, which are crystalline and composed of metal ion clusters. These clusters are coordinated with each other through different ligands, thereby resulting in the formation of a network of multidimensional structure with high surface area and porosity [9]. The integration of these materials with other class such as a hybrid structure of MOFs and CNTs, can also be harnessed for CCUS by fine-tuning the chemical functionalities. These materials can also act like selective membranes, which allow other gases to pass while trapping the CO_2 [5].

The particular class of the nanocomposites which works as a membrane for the CO_2 separation are the polymer-based nanocomposites. The permeability of these materials is increased by the incorporation of different nanoparticles, such as silica, CNTs, or other oxides in the polymer composites [10]. The incorporated nanoparticles tend to alter the networking structure by disrupting the polymer chains, thereby resulting in the enhancement of free volume of the materials, which is the responsible mechanism for the permeability enhancement. Furthermore, the fine-tuning of other properties and selectivity can also be done by tailoring the chemical nature of the nanoparticles. A particular subclass of these matrices is the biopolymer materials, which are cost-effective and eco-friendly. These biopolymers are being obtained from different biological sources such as cellulose, chitosan, and alginate, which have different functional groups like amino (-NH_2), hydroxyl (-OH) in their structure [3]. Owing to different functional groups, these materials interact through a chemisorption process with the CO_2 molecules.

Furthermore, through the incorporation of the nanofillers, the mechanical strength, stability, and sorption, the characteristics of these biopolymer materials are enhanced as they tend to increase the surface areas with porosity. Also, nowadays, hybrid nanocomposite structures are being explored to attain multiple processes involved in the technology. As an example, the hybrid nanocomposite of the MOF and graphene has been found to exhibit marvellous characteristics for being used as an adsorbent material, as the two important characteristics that are high porosity of MOF and high conductivity of graphene are now found in one structure [3, 11].

The efficiency of the nanocomposite materials for CO_2 capturing relies on different physical and chemical mechanisms at the nanoscale regime. The two fundamental mechanisms involved are physical and chemical adsorption. In the primary one, also called physisorption, the CO_2 molecules are captured on the surface through weak molecular forces [12]. The amount of CO_2 being captured depends highly on the porosity and surface area of the materials. The versatility of this process is the capability of reusing the materials for multiple cycles, as the captured CO_2 can be released from the surface either through a slight variation in pressure or temperature. The latter mechanism involves the formation of chemical bonds with CO_2, where the capturing occurs on specific sites available on the surface of the nanocomposite. This process is facilitated through functionalisation, wherein the available functional groups within the nanocomposites remain grafted on the active surface. This approach offers high selectivity for the CO_2 and remains effective even at low concentrations of CO_2. Although the bonds formed through interaction are much stronger than the latter process, they can be reversed by supplying high amounts of energy [13, 14]. Besides these two primary mechanisms, the nanocomposite materials favour the CO_2 through other approaches as well. In the case of membrane technology, the separation processes work on the principle of selective permeation. Wherein only the CO_2 is allowed to pass the desired membrane while all other gases present in the gas stream are blocked. More recent advances include the amalgamation of the capturing and the utilisation processes called the catalytic conversion mechanism. In this process, the CO_2 is not only captured but also converted into some useful fuel, such as methane, within the catalytic site. This approach proves beneficial for the feedstock by leveraging the cost of carbon capturing. The effectiveness of any mechanism relies on some of the major characteristic properties, such as available active sites, surface area, recyclability, porosity, etc. Researchers across the globe are working on achieving efficiency in different materials by fine-tuning these characteristic properties [13, 14].

Among the mentioned factor the surface area and porosity are linked together, as the surface with pores is supposed to have a high specific area. Higher surface area means higher pores, which directly increases the surface for the CO_2 adsorption. The importance of the pore size and how it affects the efficiency is discussed in another chapter. Another important parameter is fine-tuning the chemistry of nanocomposites through functionalisation of the surface with functional groups that hold a strong affinity for the CO_2, thereby helping to increase the CO_2 capturing capacity and selectivity [4, 8]. With all these material advancement's main goal is to design an integrated system with each component adding the strength, stability, selectivity, regenerability, etc., in the system. Some other aspects that need to be taken into account are the operational lifetime, rate of adsorption and disruption, which are noteworthy and need to be taken care of. For the practical implications, some of the prominent challenges are related to the large-scale production, cost-effectiveness, eco-friendliness, and recyclability. Continuous innovation and technological growth have made it possible, to some extent unlock the promises for scalable and viable technology through nanocomposite materials. These efforts are making the way for attaining the goal of zero emissions, thereby mitigating the changes in the climate and fostering the sustainable global economy [2, 15].

Looking into different engineering approaches that have been carried out in the CCUS technology through material manipulations, this chapter will enlighten the different classes of the nanocomposites. The different methods employed for designing the different nanocomposites, their thermo-mechanical properties, and the morphological aspects of these nanocomposites. Finally, the chapter will provide a brief overview of the different challenges offered in the technology at different intermediate stages.

4.2 Different Types of Nanocomposites for Carbon Capture

To address the unprecedented environmental challenges brought on by the rising levels of CO_2, material science has undergone a significant transformation. This shift has led to the development of advanced classes of materials designed specifically to tackle issues like carbon capture. Among these, nanocomposites have emerged as particularly promising, thanks to their ability to redefine the approach to CO_2 capture [2]. Nanocomposites are multi-phase materials where one phase, composed of nanoparticles, is dispersed within another bulk-phase material. Most nanocomposites consist of polymer, silica, or biopolymer matrices combined with nanofillers such as MOFs, graphene, and metal oxides, forming materials that are both highly active and processable. Among them, amine-functionalised nanocomposites have shown excellent efficiency for low-pressure CO_2 capture due to their strong chemisorptive affinity and enhanced transport properties. Similarly, MOF-based nanocomposites exhibit remarkable capture performance owing to their large surface area and tunable pore chemistry, while embedding MOFs into polymer matrices improves their mechanical stability and applicability under realistic conditions. Graphene, graphene oxides, and other carbon-based nanocomposites offer high surface areas and conductive networks that facilitate amine loading and enable Joule-heating regeneration. These materials hold great promise as membrane fillers, regenerable sorbents, and structured monoliths. In addition, other nanocomposite systems such as mixed-matrix membranes, metal oxides, and carbonates also play significant roles in carbon capture, as illustrated in **Figure 4.1**.

The strength of these materials lies in the synergistic effects between the two phases, which give rise to unique and enhanced properties not found in their individual components. These enhancements are largely made possible by nanoengineering, which allows for precise control over the structural and chemical characteristics of the material. As a result, researchers have been able to create adsorbents and membranes that offer exceptional features, such as high selectivity, large surface area, rapid reaction kinetics, and long-term stability [16]. The wide range of available nanoparticles and the flexibility in choosing a host matrix have led to the creation of diverse families of nanocomposites, each with unique advantages for CCUS applications [17].

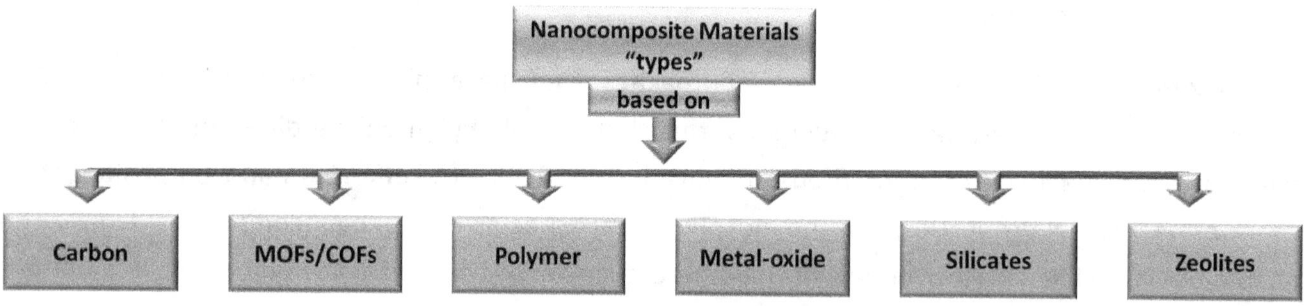

Figure 4.1: Schematic overview representing the different classes of nanocomposite materials employed for the capturing of CO_2 through different processes.

In the following section, we will provide an overview of the various nanocomposites that have shown potential in CO_2 capture. This will include their composition, mechanisms of action, along with their respective benefits and limitations.

Carbon-Based Nanocomposites: Among the various types of nanocomposites, this particular class stands out as one of the earliest and most fundamental. These nanocomposites are developed using different carbon allotropes, such as graphene and carbon nanotubes (CNTs), which serve either as adsorbents or as nanofillers within a host matrix **(Figure 4.2)** [18]. What makes these materials especially valuable is not only their outstanding inherent properties, but also their ability to be chemically functionalised, allowing researchers to tailor their performance for specific applications [9, 19].

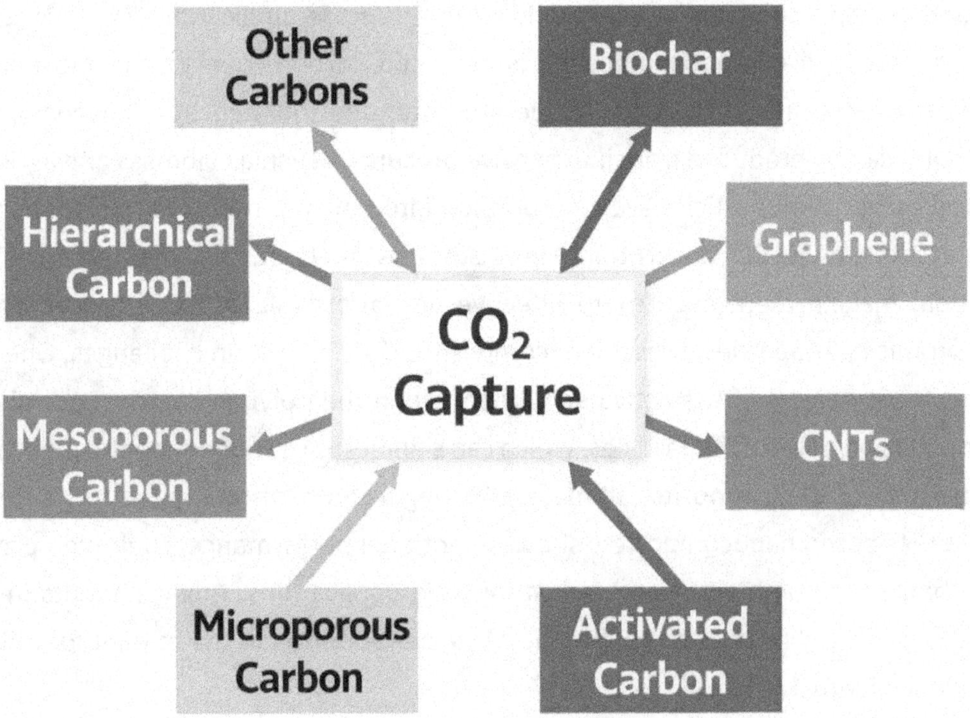

Figure 4.2: An overview of different carbon-based materials employed for CO_2 capturing.

Among the carbon allotropes, graphene is especially notable. It is a remarkable single-layer material composed of sp^2-hybridised carbon atoms arranged in a honeycomb-like structure. CNTs, on the other hand, can be thought of as rolled-up sheets of graphene. Both structures exhibit an exceptionally high surface area and are capable of adsorbing CO_2 efficiently through physical interactions such as van der Waals forces [20]. However, their full potential is realised when they are integrated into polymer matrices to form composite materials. When used in nanocomposite membranes, these carbon nanostructures significantly enhance performance by contributing to a high surface area, improved adsorption capacity, and excellent selectivity for CO_2. This leads to more efficient carbon capture and greatly improved membrane properties overall.

Furthermore, the efficiency of these materials can be significantly enhanced through chemical functionalisation. This typically involves attaching functional groups to the surface of the material, specifically those with a strong affinity for CO_2 [21]. Among the wide range of possible functional groups, amine functionalisation has proven to be one of the most effective approaches for CCUS applications. The amine group ($-NH_2$) has a strong tendency to react with CO_2 through a process called chemisorption, where chemical bonds are formed, leading to the creation of carbamate species. Because chemisorption involves stronger and more selective interactions than physical adsorption (physisorption), it greatly increases the material's CO_2 capture capability. What makes this approach particularly valuable is its effectiveness even under challenging conditions, such as low CO_2 partial pressures, which are typical in post-combustion scenarios or even in ambient air. This makes amine-functionalised nanocomposites a promising and versatile solution for real-world carbon capture applications [22]. Another widely used nanomaterial in the field of nanocomposites is activated carbon, one of the most traditional and accessible options. It has attracted considerable attention due to its cost-effectiveness, high porosity, and the fact that it can be produced from inexpensive precursors such as biomass and agricultural waste. When activated carbon nanoparticles are incorporated into polymer matrices to form nanocomposites, they serve as active sites for CO_2 adsorption. These surfaces can be further optimised through chemical functionalisation, making them even more effective and affordable as CO_2 adsorbents [23]. However, despite their promising properties, these nanocomposites do face certain challenges. One major issue is achieving uniform dispersion of the activated carbon within the polymer matrix. Poor dispersion often leads to agglomeration of the nanoparticles, which can significantly reduce both the mechanical strength and the surface area of the composite, ultimately limiting its adsorption capacity. As a final note, while graphene- and CNT-based nanocomposites show exceptional performance, their high production costs remain a key barrier to commercial and industrial-scale applications. Advancements in cost-effective synthesis methods are still needed to enable their broader adoption in the market for different carbon-based adsorbents **(Figure 4.3)** [15].

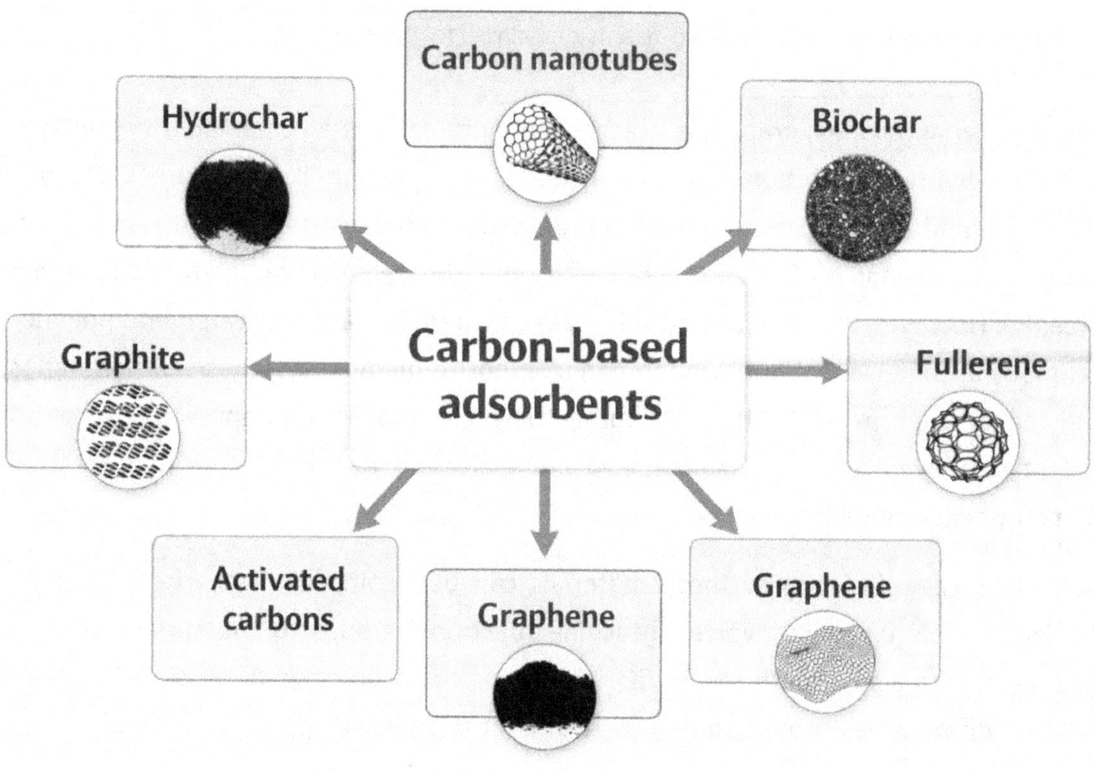

Figure 4.3: Different carbon-based adsorbents used for the CCUS applications. Copyright Credit: (Ref. [15], Materials Today Sustainability, Elsevier, 2023)

MOF-Based Nanocomposites: MOFs are a fascinating class of materials known for their highly crystalline structure and exceptional porosity. Owing to their unique features, MOFs have garnered significant attention and established themselves as promising candidates for CCUS applications. MOFs are composed of metal ions or metal clusters connected by organic linkers, forming multidimensional crystal structures [16]. A key advantage of MOFs is the flexibility in design. Researchers can precisely select the metal centres and organic molecules during synthesis to fine-tune the material's properties. This level of control has enabled the development of MOFs with extremely high surface areas, reaching up to 7,000 m^2/g, along with tailored pore sizes, shapes, and chemical environments. Despite their excellent CO_2 adsorption potential, pristine MOFs do face some practical limitations. These include poor mechanical strength, processability, and instability in humid or acidic environments. To overcome these drawbacks, MOFs are increasingly being integrated into nanocomposite structures [7, 24].

One of the most effective strategies in this regard is the development of Mixed-Matrix Membranes (MMMs) [21]. In MMMs, MOF nanoparticles are uniformly dispersed within a polymer matrix. This hybrid structure combines the strong CO_2 affinity of MOFs with the mechanical robustness, flexibility, and ease of processing offered by polymers. However, the performance of MMMs. heavily depends on the quality of interaction between the MOF particles and the polymer matrix. Weak adhesion can lead to the formation of non-selective interfacial voids, which reduce membrane efficiency. Therefore, optimising the interface between the MOF and the polymer is crucial in maximising the benefits of these advanced composite materials [25]. In addition to the development of MMMs, other innovative strategies have also been explored for the development of robust MOF-based hybrid nanocomposites, notably combinations like MOF-graphene and MOF-CNT structures. These hybrid materials offer enhanced mechanical strength, improved electrical conductivity, and greater durability and handling stability, making them highly attractive for practical applications. What sets these nanocomposites apart is the ability to finely tune the properties of the MOF component. By carefully selecting the metal ions, organic linkers, and framework geometry, researchers have successfully engineered MOF-based composites with optimal pore sizes and high selectivity for CO_2 capture. Moreover, the presence of organic linkers opens up additional opportunities for chemical functionalisation, further boosting the material's affinity for CO_2. The overview of progress made by different MOFs in the due course of time is represented in **Figure 4.4** [26]. Overall, the versatility in design and performance tuning of MOF-based nanocomposites is increasingly seen as front-runners in meeting the growing demands of CCUS technologies. Their customisable nature and high performance make them a promising solution for the next generation of carbon capture materials.

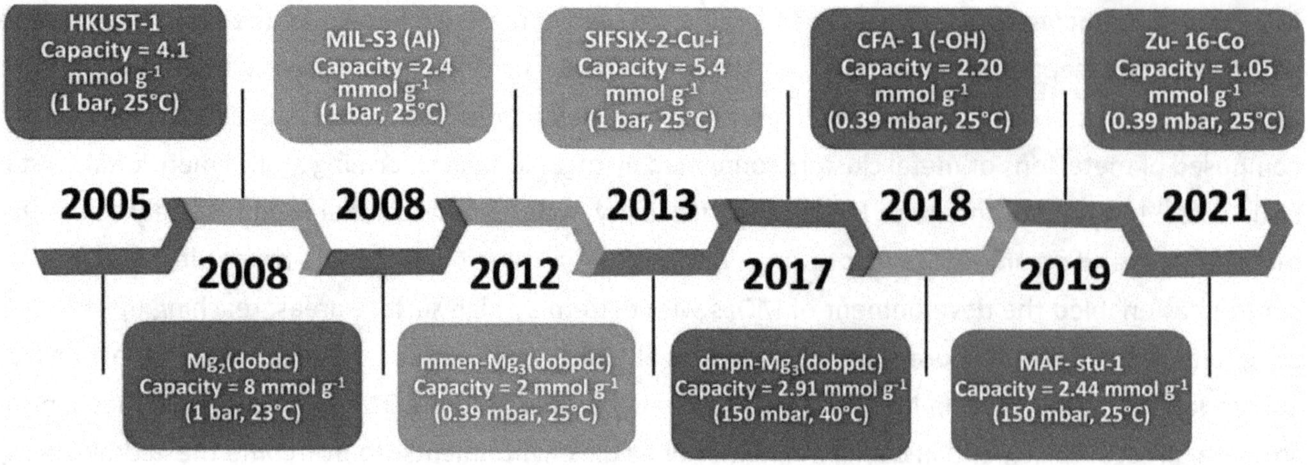

Figure 4.4: Progress made in the MOFs for the CO_2 capturing starting from the HKUST-1 in 2005, Mg_2 (dobdc), and MIL-53 in 2008, mmen-Mg_2 (dobpdc) in 2012, SIFSIX-2-Cu-I in 2013, followed by successive progress in years with Zu-16-Co in 2021. Copyright Credit: (Ref. [26],Journal of Environmental Chemical Engineering, Elsevier , 2022)

Polymer-Based Nanocomposites: These nanocomposites, generally used as membranes, represent a scalable, energy-efficient, and cost-effective solution for CO_2 capture, making them one of the most promising approaches available today. These membranes work by separating gases through a process called the solution-diffusion mechanism. Essentially, gases pass through the membrane based on two key properties: permeability, how easily the gas passes through, and selectivity, how well the membrane distinguishes CO_2 from other gases. Together, these properties define the performance limit known as the Robeson upper bound [27]. For an ideal membrane, it is necessary to have both high permeability and high selectivity, but achieving this balance is very challenging. These two properties often behave like uncertainty conjugate variables: improving one tends to reduce the other, creating a trade-off that limits overall efficiency [28]. Fortunately, this limitation can be addressed by adding nanofillers to the polymer matrix to create nanocomposite membranes. These tiny fillers interact with the polymer chains, altering the membrane's internal structure by increasing the free volume, essentially creating more space for gas molecules to move through. This increase in free volume boosts gas permeability. At the same time, nanofillers enhance selectivity in several ways. They can slow down larger gas molecules by creating more complex, winding paths through the membrane.

Additionally, nanofillers can be chemically functionalised, for example, by adding polar groups that specifically attract CO_2 molecules, which are polar, more than non-polar gases like N_2 or CH_4. This selective interaction improves the membrane's ability to separate CO_2 more efficiently [29]. Building on these advances, a new generation of biopolymer-based nanocomposites has emerged. Made from natural materials like cellulose, chitosan, starch, and alginate, these membranes not only offer excellent performance but are also biodegradable and eco-friendly. The natural functional groups (like hydroxyl and amine groups) present in these materials provide a strong affinity for CO_2, thereby enhancing capture efficiency. Owing to their sustainable origins and promising capabilities, biopolymer-based nanocomposites could play a key role in the future of carbon capture technologies, offering an eco-friendly, responsible path forward [30].

Metal-Oxide-Based Nanocomposites: These nanocomposite materials are primarily designed for CO_2 capture under high-temperature conditions, such as those encountered in chemical-looping combustion and sorption-enhanced reforming processes. Their working principle is based on reaction reversibility [31]. In this process, CO_2 reacts with a metal oxide to form a stable metal carbonate. To release the captured CO_2, the carbonate is subjected to a temperature higher than its formation temperature, triggering the reverse reaction and liberating CO_2. However, using pure metal oxides such as CaO or MgO poses challenges. They are not only expensive but also prone to structural deterioration during repeated carbonation-decarbonation cycles [32]. With each cycle, the affinity of these oxides for CO_2 decreases due to sintering and loss of active surface area. A promising solution lies in the development of nanocomposite materials. This is typically achieved by dispersing metal oxide nanoparticles onto stable, high surface area supports such as alumina, zirconia, or even CNTs. These supports not only provide structural stability but also prevent particle agglomeration at elevated temperatures. As a result, the nanocomposites maintain both high-temperature stability and strong CO_2 affinity over multiple cycles. Most importantly, this approach is cost-effective, making it a versatile and practical pathway for efficient CO_2 capture under industrial conditions [33].

Inorganic Nanocomposites: In order to design materials that can perform effectively in the extreme conditions often encountered in CCUS applications, researchers have explored a variety of options. One promising group is inorganic nanocomposites, which are especially well-suited for capturing CO_2 at high temperatures and in challenging atmospheric conditions. Among these, silica- and zeolite-based inorganic materials have stood out for their excellent stability and strong CO_2 capture performance [34]. A brief overview of these materials and their advantages is provided below:

Silica-based Nanocomposites: These nanocomposite materials are derived from mesoporous silica, which is well known for its exceptionally high surface area. The pore size of mesoporous silica typically ranges from 2 to 50 nm. Materials such as MCM-41 and SBA-15 are widely recognised for their uniform pore structure and remarkably large surface areas, sometimes reaching up to 1000 m2/g. Silica, being robust in nature, exhibits excellent thermal and chemical stability. However, in its natural state, silica shows very low affinity for CO_2 capture. This limitation can be overcome through functionalisation, which significantly enhances its performance. The key to this functionalisation lies in the abundant surface -Si-OH groups, which serve as active sites for grafting functional molecules. Among the most effective molecules for this purpose are alkylamines [35]. Once incorporated, these groups transform silica from a mere physical adsorbent into a chemical adsorbent. This functionalisation not only improves the selectivity of the material but also boosts its efficiency, even under humid atmospheric conditions.

Furthermore, the CO_2-capturing capacity of functionalised silica can be fine-tuned by regulating pore size, adjusting the number of alkylamine groups introduced, or modifying the chemical structure of the amine itself [15]. Different strategies to be employed for future progress to be made in the silica-based nanocomposite are depicted in **Figure 4.5**.

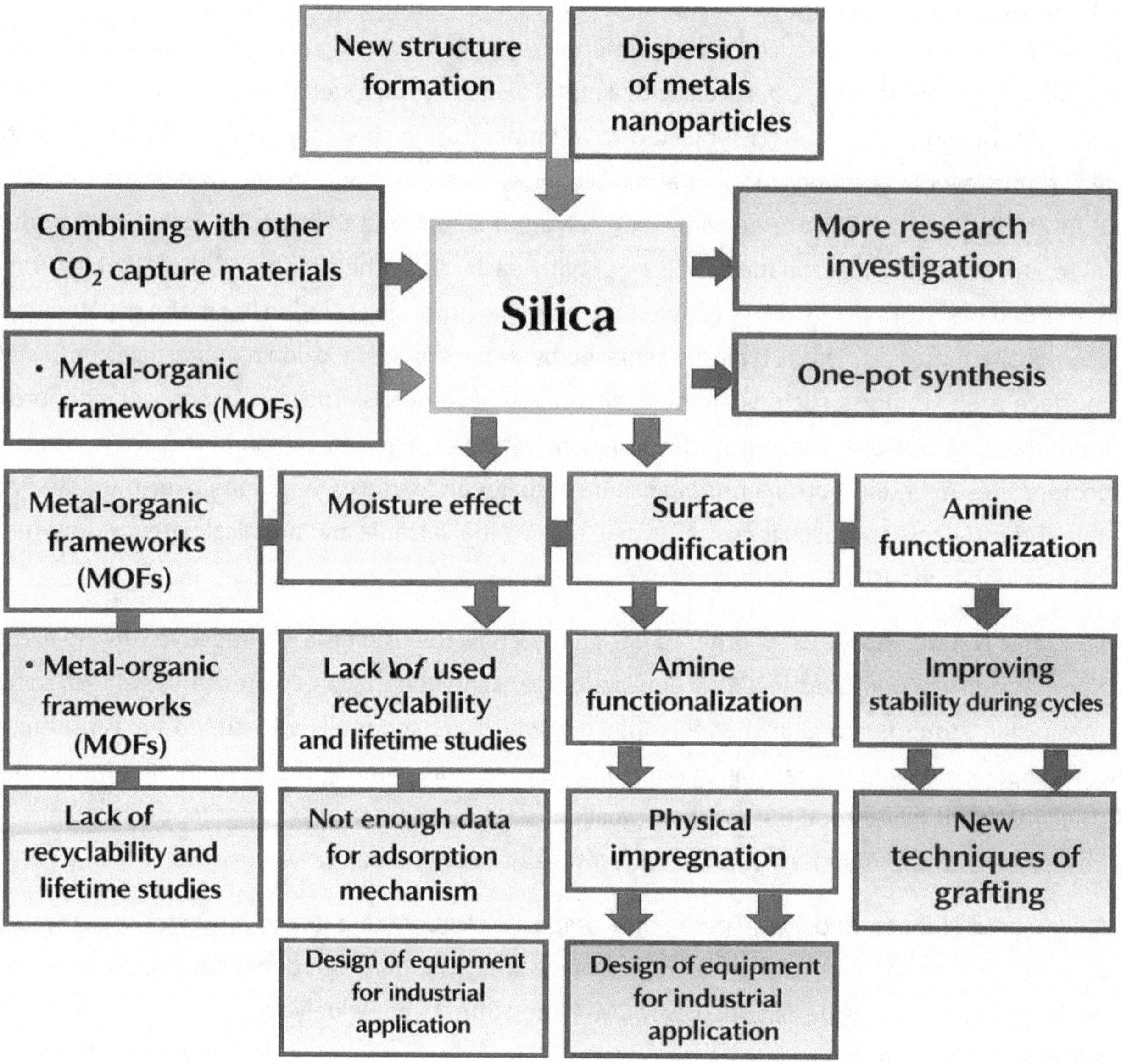

Figure 4.5: Different strategies to be employed for the success of the silica-based adsorbents. Copyright Credit: (Ref [15], Materials Today Sustainability, Elsevier, 2023)

Zeolite-based Nanocomposites: Zeolites are rigid, three-dimensional network structures characterised by their uniform pore sizes, typically smaller than 2 nm. They are crystalline aluminosilicate minerals, well known for their precise pore distribution, which has made them invaluable in catalysis and separation processes. One of their most remarkable features is the presence of strong electrostatic fields within the pores, which significantly enhances their ability to separate gases. Owing to this unique structural configuration, zeolites exhibit a strong affinity for selectively capturing CO_2 from gas mixtures [36]. These materials have proven effective, especially for post-combustion CO_2 capture and natural gas sweetening. However, their small pore size also poses a challenge, that is, slow kinetic diffusion, which limits the rate of CO_2 adsorption and release. To overcome this limitation, researchers have turned to zeolite-based nanocomposite materials. There are two main approaches to achieve this: (1) synthesising zeolite nanocrystals and dispersing them into polymer membranes to form MMMs. or (2) integrating microporous zeolites with mesoporous materials. Both strategies reduce the diffusion pathway, thereby enhancing the kinetics of adsorption and desorption [37].

4.3 Different Approaches for the Fabrication of Nanocomposites for Carbon Capture Applications

The success of nanocomposites for CCUS applications largely depends on advanced material engineering strategies. In these approaches, it is the design and fabrication of the composite, rather than the performance of the individual constituents, that ultimately determines its effectiveness. The way nanoscale properties are translated into the macroscopic regime is strongly influenced by the chosen synthetic method. Key material properties, such as structure, pore size, distribution of nanofillers, phase, and morphology, are all dictated by the synthesis approach. These characteristics, in turn, directly affect critical parameters, including CO_2 affinity, selectivity, kinetics, stability, and overall performance [38]. Therefore, the efficiency of a nanocomposite for CCUS hinges on both the science and the art of synthesis. Careful selection of the fabrication route is essential for developing highly efficient and cost-effective materials. Broadly, synthesis methods can be categorised into two main types: in-situ and ex-situ, which are outlined below:

In-Situ Synthesis: The "one-pot" synthesis approach allows nanoparticles to grow directly within the host matrix. Because the nanoparticles develop through an in-situ approach, it proves to be efficient, as this method ensures a uniform distribution and strong bonding between the filler and the matrix [39]. For example, in the fabrication of MOF-based nanocomposites, both the metal and organic precursors are directly introduced into a polymer solution.

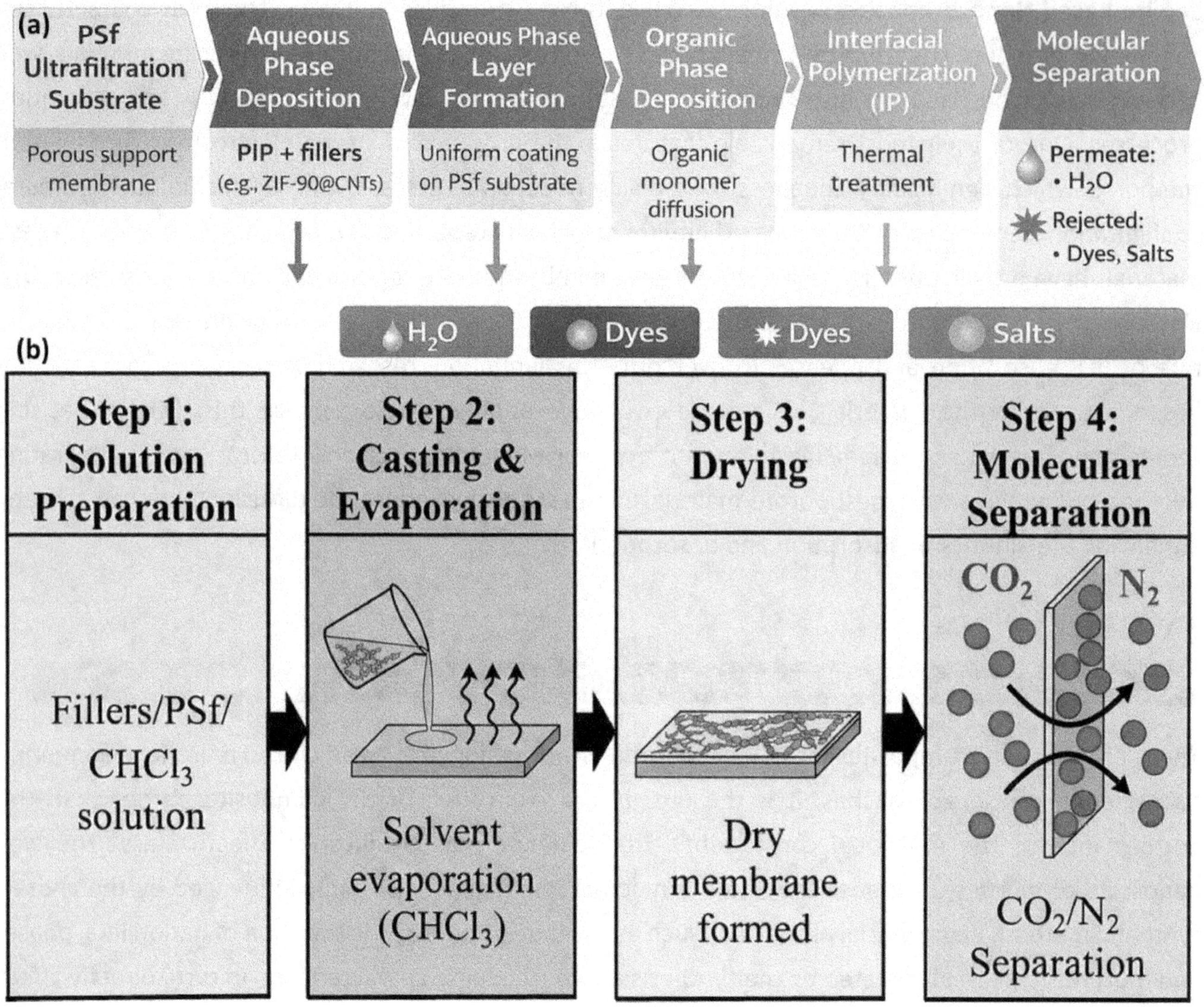

Figure 4.6: Pictorial representation for synthesis of MOF@CNTs membranes through (a) nanofiltration, (b) gas separation, for molecular separation. Copyright Credit: (Ref. [40], RSC Applied Interfaces, RSC, 2025)

As the process proceeds, MOF particles nucleate and become embedded within the polymer chains. Alternatively, if nanoparticles are already available, a monomer can be polymerised in their presence, leading to their incorporation during matrix formation. The most significant advantage of this method is the creation of structures with strong interactions between the nanofillers and the polymer matrix. This close bonding reduces defects and enhances synergistic effects, making such composites efficient and particularly effective in applications like gas separation membranes **(Figure 4.6)** [40].

Ex-situ Synthesis. This is one of the most widely used approaches for synthesising nanocomposites. In this method, the nanofillers and the matrix are prepared separately using their respective synthesis routes. Once both components are obtained, they are combined through blending, mixing, or similar techniques to form the final nanocomposite. For example, pre-synthesised carbon CNTs can be incorporated into molten polymer to fabricate the desired composite. This strategy allows precise control over the synthesis of each component, and consequently, their individual properties. However, a key challenge lies in achieving uniform dispersion of the nanofillers within the matrix. Materials produced this way often suffer from nanoparticle agglomeration, which negatively impacts performance. Another critical aspect is ensuring good interfacial compatibility between the nanofillers and the matrix. Without this, defects can form at the interface, ultimately reducing the efficiency of the composite material [41]. Some of the schematics employed for synthesising different materials, like polymer adsorbents, MOFs, and other multidimensional materials, are presented in **Figure 4.7** [8, 41]. Within the broader framework of material design, several specialised and highly precise techniques have been developed for fabricating different types of nanocomposites. Each class of materials, such as carbon-based systems, MOFs, and polymers, etc., has its own tailored approach, with careful attention given to each step. This not only ensures a deeper understanding of the underlying chemistry but also provides ways to fine-tune the material properties for specific applications. Some of these approaches are discussed below:

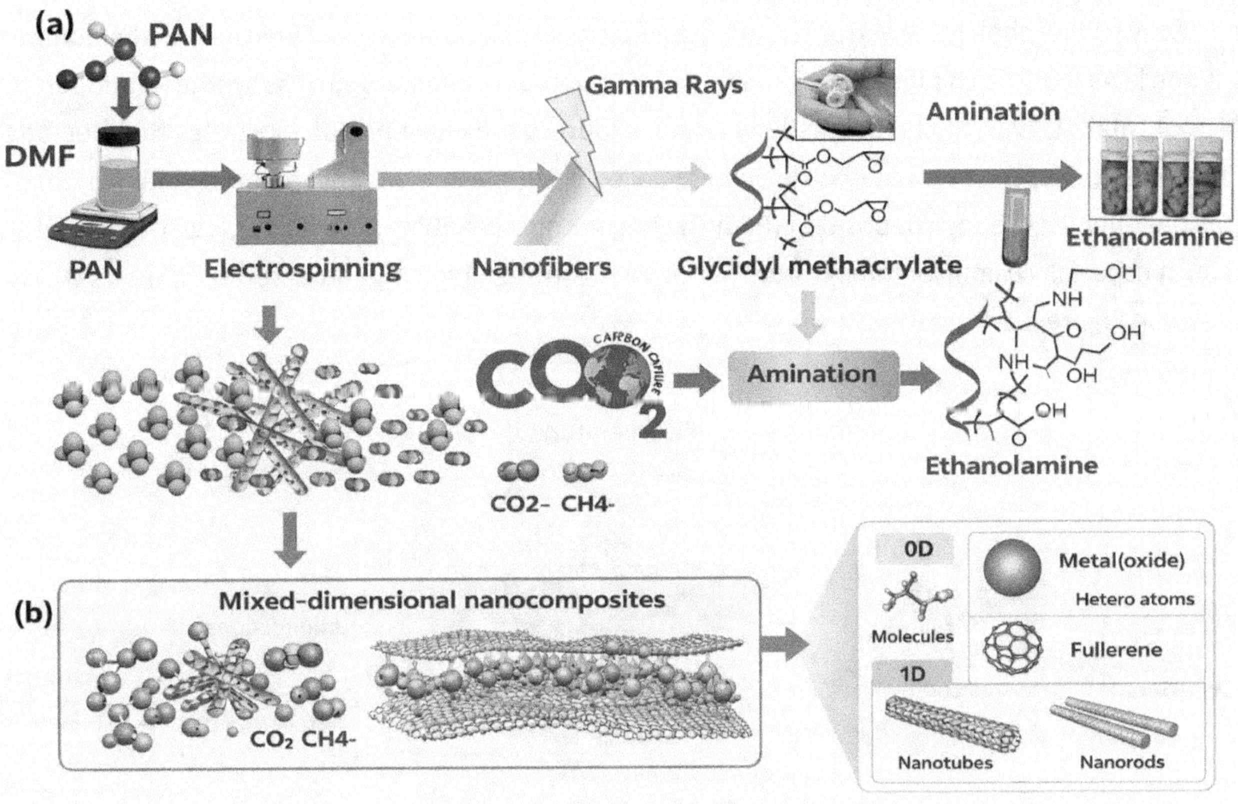

Figure 4.7: Schematic representation for different ex-situ approaches (a) for preparation of polymeric adsorbents, (b) different multidimensional materials through the combination of 2D and other mixed-dimensional composites, for the CO_2 capturing and storage. Copyright Credit: (Ref. [41], Scientific Reports, Nature, 2023, and Ref.[8], npj 2D Materials and Applications, Nature, 2023)

4.4 Synthesis of Carbon-based Nanocomposites

The development of these nanocomposites generally involves two main stages: first, the synthesis of carbon nanofillers such as graphene, CNTs, and other allotropes; and second, their integration into a suitable polymer host matrix. Each of these stages requires distinct synthesis approaches, which are outlined below.

4.4.1 Synthesis of Carbon Nanofillers

Chemical Vapour Deposition (CVD): This is one of the most prominent approaches for the large-scale synthesis of CNTs and graphene sheets. It is widely accepted and has already reached the extent of commercial use. The process is carried out in a high-temperature furnace, where hydrocarbons such as methane, ethylene, or acetylene are decomposed. In the initial step, a catalyst, commonly iron, nickel, or cobalt, is introduced onto the substrate, either as thin films or as nanoparticles. Then the substrate is heated inside the furnace, the hydrocarbon decomposes, and thereby the elemental carbon gets dissolved into the catalyst. Once the catalyst becomes saturated, carbon begins to precipitate out, leading to the growth of either graphene sheets or CNT structures [42]. The properties of the final product, particularly its morphology, are strongly influenced by synthesis parameters. For instance, the number of graphene layers, or the chirality and diameter of CNTs, depend on factors such as synthesis temperature, choice of catalyst, type of gas, and its flow rate. To address the challenge of substrates that degrade at high temperatures, the method was further advanced using plasma enhancement, resulting in Plasma-Enhanced Chemical Vapour Deposition (PECVD). This variation enables synthesis at lower temperatures, making it especially useful for temperature-sensitive substrates. Typical general setup for the CVD system is shown in **Figure 4.8** [42].

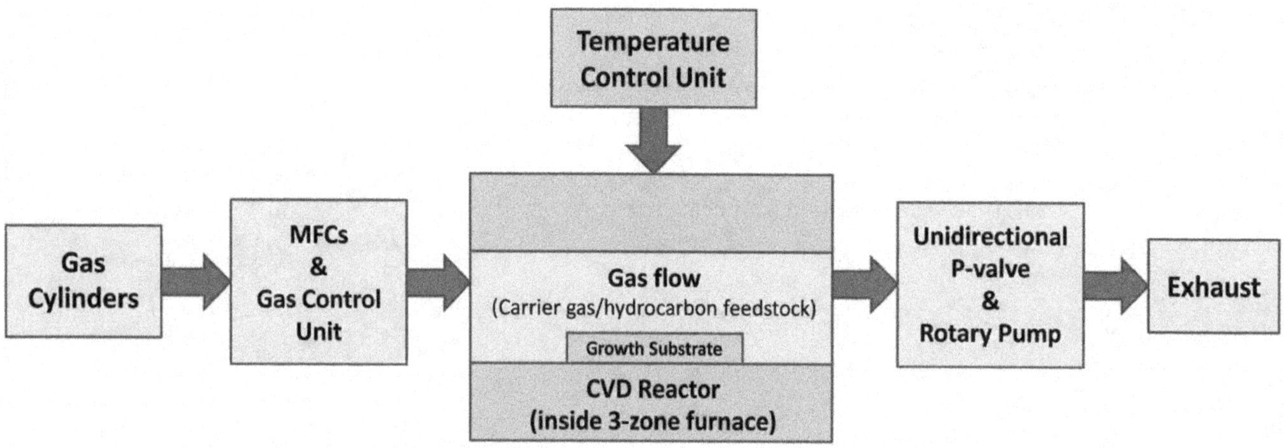

Figure 4.8: Pictorial representation of the CVD system with other important accessories.

Hydrothermal Carbonation (HCT): This method is widely recognised for its eco-friendly nature, as it relies on renewable raw materials such as biomass and agricultural waste to produce carbon-based nanoparticles. Because it employs natural precursors, it is often referred to as a green synthesis approach. In this process, biomass rich in carbohydrates is mixed with water and heated in an autoclave at around 200–250 °C, generating autogenous pressure. Under these conditions, the precursors undergo dehydration and polymerisation reactions, which lead to the formation of carbonaceous nanostructures [43]. These structures often appear in an amorphous form with a core-shell architecture and are typically subjected to calcination to achieve proper crystallisation. The resulting materials retain a surface rich in naturally attached functional groups. These groups enhance the affinity of the materials for CO_2 and also improve their solubility in polar solvents, thereby boosting their adsorption capacity [44].

Overall, this one-step, low-temperature HTC method provides a cost-effective and sustainable route for converting biomass waste into valuable carbon materials with broad application potential. The schematic approach depicting the process of synthesising the graphene-based composites is presented in **Figure 4.9** [45].

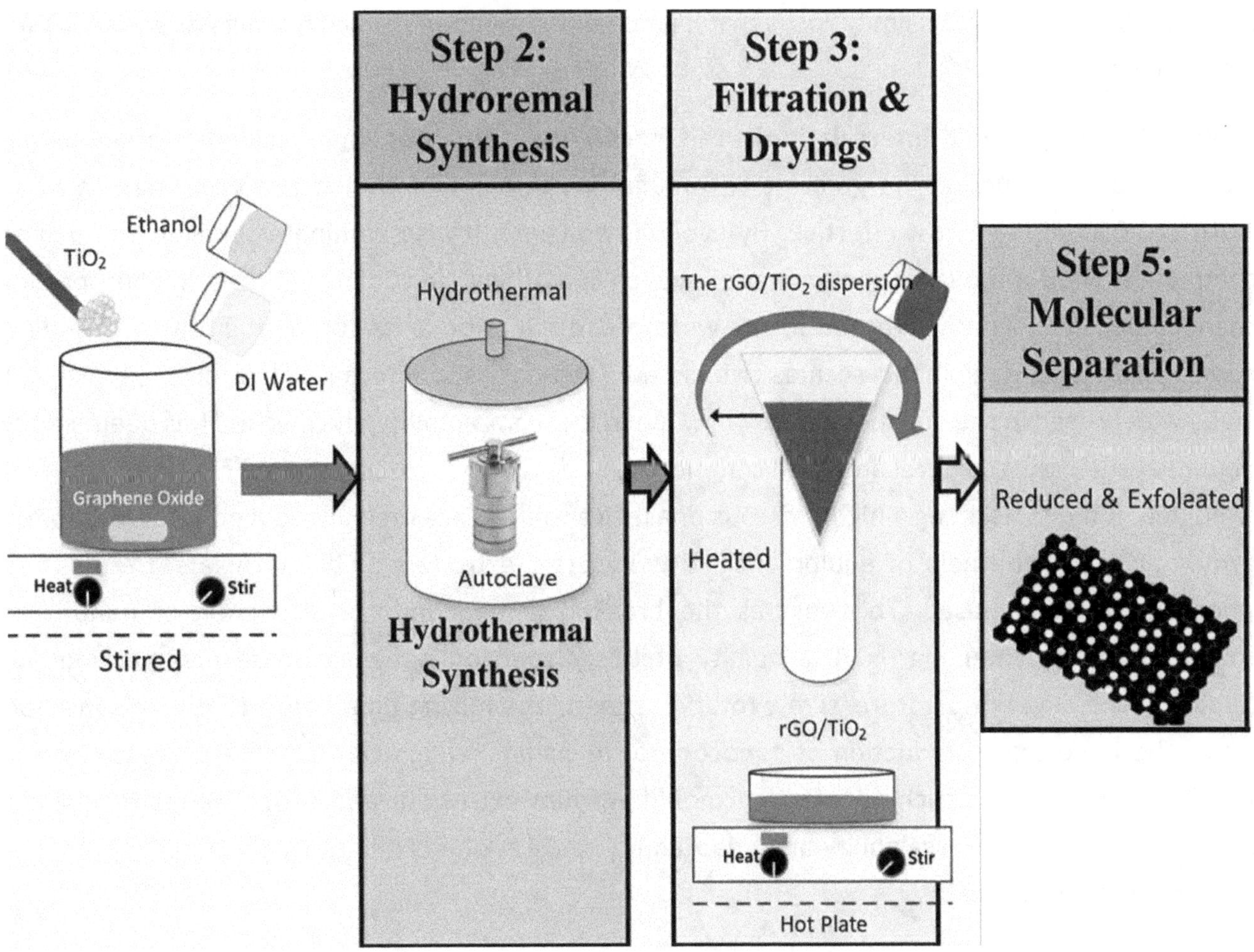

Figure 4.9: Steps involved in the hydrothermal approach for synthesising the graphene-based nanocomposites while using TiO_2. Copyright Credit: (Ref. [45], ACS Omega, ACS 2024)

4.4.2 Integration into Nanocomposite Structures (Ex-situ)

After successful synthesis of the nanoparticles, the next step is their integration into the polymer matrix. This process is being carried out through one of the following, briefly explained ex-situ techniques:

Solution Mixing and Casting: This is one of the simplest and most widely used approaches for synthesising nanocomposite materials. In the first step, pre-synthesised nanoparticles such as CNTs or graphene are dispersed in a suitable solvent. Since these nanoparticles are strongly bound together by van der Waals forces, achieving a uniform and stable dispersion is often challenging. To overcome this, techniques such as strong sonication or high-shear mixing are typically employed. To further improve dispersion, enhance compatibility, and prevent re-agglomeration, surface functionalisation of the nanoparticles is often carried out.

Meanwhile, the selected polymer is dissolved in the same or in a miscible solvent. The two prepared solutions are then combined to form a homogeneous suspension, which is cast onto a substrate in what is known as the casting process. The coated substrate is subsequently heated to evaporate the solvent, leaving behind a thin film of nanocomposite. While this method is cost-effective and straightforward, its main limitation is that it is solvent-dependent, which restricts its scalability for large-scale production [46].

Melt Blending (Melt Intercalation): This technique serves as the backbone of industrial nanocomposite production. Its greatest advantage lies in the fact that it is solvent-free, making it both eco-friendly and cost-effective. By avoiding solvents, it also eliminates challenges such as incomplete solvent removal or contamination, which can otherwise compromise the final properties of nanocomposites. Another reason for its widespread adoption is its compatibility with standard polymer processing equipment, such as twin-screw extruders and internal batch mixers, which are already widely used in the polymer industry [47]. Due to this compatibility, the method has been readily accepted at the industrial level and even commercial scale. In this process, the polymer matrix is first heated until it melts, forming a highly viscous phase. Nanofillers are then introduced into this molten polymer, and with the help of equipment like twin-screw extruders or batch mixers, the mixture is thoroughly homogenised. This ensures the breakdown and uniform dispersion of nanofillers throughout the polymer matrix. The quality of this dispersion depends on several parameters, including processing temperature, screw rotation speed, and mixing time. Importantly, this method enables the large-scale production of nanocomposite pellets, which can later be processed using conventional techniques such as injection moulding, film extrusion, and other standard polymer fabrication methods. This scalability and adaptability make it one of the most practical approaches for industrial applications [48].

4.5 Synthesis of MOF-Based Nanocomposites

The synthesis of these MOF nanocomposites is one of the challenging tasks. Properly acquiring the desired product, it requires proper control over the crystallisation process during the integration. As it should be done in a proper way without disturbing either of the two structures. The synthesis of these nanocomposites is being acquired through either of the following, briefly explained approaches:

In-situ Solvothermal Synthesis: This is the primary approach employed for the synthesis of MOF. The nanocomposites acquired through this approach are believed to have strong interfacial bonding. In this process, the main steps for the synthesis of the desired MOF nanocomposite are: four precursors are mixed together and then sealed in the Teflon-coated autoclave. The precursors include: appropriate metal salt (zinc/copper nitrate), organic linker (terephthalic acid, 2-methylimidazole), appropriate solvent (DMF, ethanol), and nanofillers. The sealed autoclave is then subjected to an appropriate temperature ranging between 80 and 200°C, for a time period which ranges from a few hours to days. If the chosen solvent in the process is water, then it is called hydrothermal. Under proper temperature conditions, the nucleation of MOF takes place directly on the required surface or the host matrix, thus considered as in-situ in nature. The in-situ synthesis results in the development of efficient chemical and physical linkages in the materials. Besides that, it favours the minimisation of defects and enhances the transport pathways in the materials, thereby making it suitable for the CCUS applications [49]. Similarly, employing the wood-based templated, researchers have also developed the MOFs, as illustrated in **Figure 4.10**, along with the mechanism of the dissolution-regeneration [50].

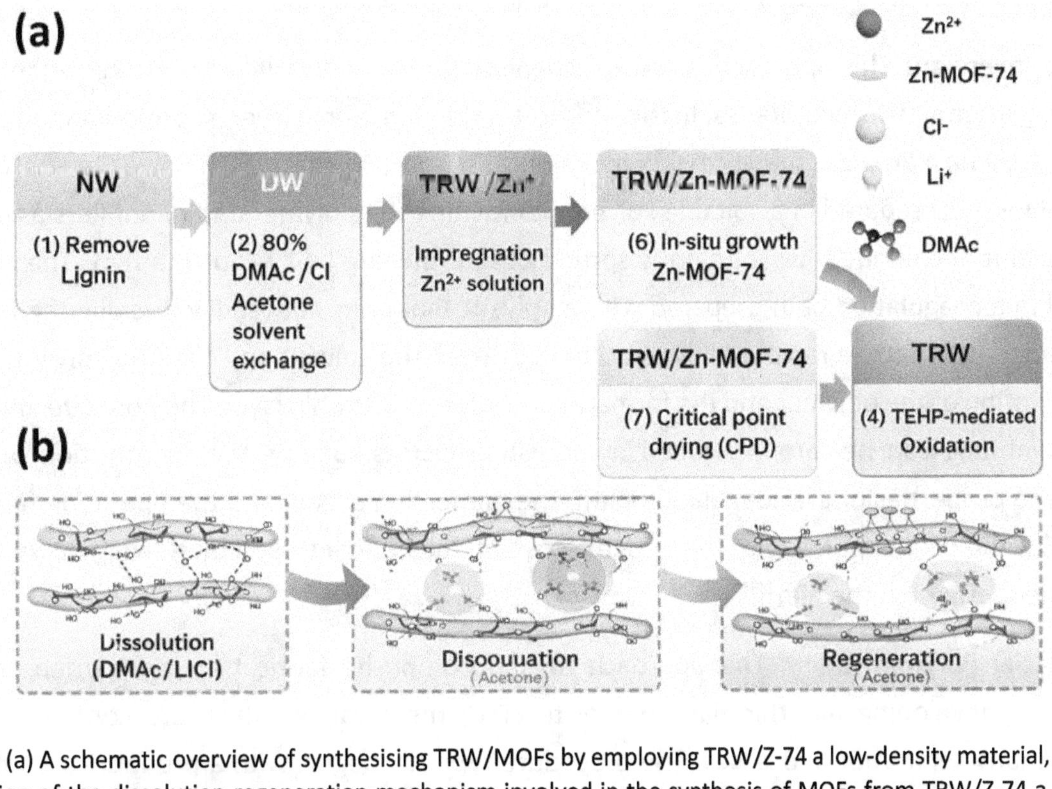

Figure 4.10: (a) A schematic overview of synthesising TRW/MOFs by employing TRW/Z-74 a low-density material, (b) Pictorial representation of the dissolution-regeneration mechanism involved in the synthesis of MOFs from TRW/Z-74 a low-density material. *Copyright Credit: (Ref. [50] ,Journal of Materials Chemistry A, RSC, 2023)*

Microwave-Assisted Synthesis: This approach is similar to the above-discussed one, with the only variation that, in this case, the heating approach is changed from conventional to microwave. Using this rapid and uniform approach for the heating decreases the synthesis time from hours to minutes and days to hours, respectively. Besides this, the other advantages associated with this approach are the development of nano-MOF structures, which are beneficial for developing the nanocomposite with a density of fillers [51].

Layer-by-Layer (LBL) Assembly: This method of synthesising MOF-based nanocomposites is particularly valuable for applications that demand ultrathin or precisely controlled films. It provides exceptional accuracy and reliability in designing materials for such purposes. The process follows a bottom-up strategy, where layers are built sequentially on a substrate. For example, the substrate is first immersed in a metal ion solution, followed by immersion in an organic linker solution. Repeating this cycle leads to the growth of a highly oriented crystalline MOF film. In this way, it becomes possible to tune the film thickness down to the angstrom scale finely. Beyond thin films, this approach can also be extended to fabricate other MOF-based nanostructures by selecting appropriate deposition solutions [52].

4.5.1 Synthesis of Polymer-Based Nanocomposite Membranes:

The synthesis of nanocomposite membranes plays a vital role in gas separation applications. These membranes are designed to be thin, defect-free, and highly efficient in carrying out separation processes. The key approaches used for their fabrication are briefly outlined below:

Phase Inversion: This approach, widely recognised at the industrial level, is commonly used for synthesising asymmetric membranes. In this method, a thin functional layer is formed on top of a porous substructure, which provides the necessary mechanical support. The process involves several steps. First, a dope solution is prepared in a suitable solvent, containing the polymer and nanofillers as precursors. Using a casting technique, this solution is spread onto a moving belt to form a film. The film is then immersed in a coagulation bath prepared with a solvent that does not readily dissolve the polymer. As the film enters the bath, a rapid exchange occurs between the solvent and the non-solvent, leading to the precipitation of the polymer and the formation of a dense, ultrathin layer. The pore size and thickness of the membrane can be carefully tuned by adjusting factors such as the composition of the dope solution, the properties of the coagulation bath, the immersion speed, and the operating temperature. Additionally, the overall porosity of the structure can be further controlled by slowing down the rate of diffusion deeper within the film [53].

Interfacial Polymerisation: This approach to nanocomposite fabrication is regarded as the **gold standard** for developing ultrathin-film composite (TFC) membranes, which are considered state-of-the-art in many separation processes. The procedure typically begins with a pre-fabricated membrane support, such as polysulfone. Before the reaction, the support is saturated with an aqueous monomer solution, usually containing an amine. Once removed, the support is carefully drained to eliminate any excess solution from the surface. In the next step, the saturated support is exposed to an organic solution containing a highly reactive monomer, such as acyl chloride. The moment aqueous and organic phases come into contact, polymerisation occurs instantly at the liquid-liquid interface. This rapid reaction produces a strongly cross-linked, dense, and ultrathin polyamide selective layer. When fabricating nanocomposite TFC membranes, nanofillers are pre-dispersed into one of the monomer solutions before the interfacial polymerisation step. During the reaction, these nanofillers become embedded within the polyamide layer, resulting in enhanced performance by combining the selectivity of the polyamide with the functional advantages of the nanomaterials [54].

4.5.2 Synthesis of Inorganic and Metal Oxide Nanocomposites

These nanocomposites exhibit excellent stability and a high surface area, along with a large number of active sites, making them highly suitable for CO_2 adsorption [55]. They can be synthesised using a variety of methods, which are briefly outlined below:

Sol-Gel Process: This is a wet chemical approach commonly used for synthesising inorganic materials and their composites. The most frequently employed precursors in this process are metal alkoxides. The synthesis typically occurs in two main steps: hydrolysis followed by condensation of molecular polymers. Taking silica as an example, the hydrolysis of tetraethyl orthosilicate (TEOS) produces silicic acid in the presence of water and a catalyst, which can be either acidic or basic. The resulting silicic acid undergoes condensation, leading to the formation of a Si-O-Si bonded network and thereby producing a colloidal solution known as a sol. As the bonds interconnect further, the sol gradually transitions into a solid three-dimensional network, forming a gel. To develop highly ordered mesoporous materials such as MCM-41 or SBA-15, specific templating agents are added to the precursor solution. These templates are often long-chain cationic surfactants, which self-assemble into cylindrical micelles. The silica framework then grows around these micelles, eventually forming ordered micellar rods. In the final stage, the substrate and excess solvent are removed through calcination, yielding a silica structure with uniform, hexagonally packed pores. For in-situ synthesis of nanocomposites, secondary metal precursors can be introduced during the early sol formation phase, enabling their integration into the final mesoporous structure [56].

Grafting: This is an ex-situ approach used for chemically modifying porous structures and is therefore often referred to as post-synthesis functionalisation. A common application of this method is the preparation of amine-functionalised adsorbents. In this process, a material with a high surface area is selected and then functionalised by reacting it with molecules that possess two key features: (i) the ability to bond with the surface chemically and (ii) a functional group that enhances CO_2 capture. For instance, in the case of silica, the functional molecule most often employed is (3-aminopropyl) triethoxysilane (APTES). APTES reacts with the silanol groups present on the silica surface, forming Si-O-Si bonds. This reaction effectively grafts amine-terminated propyl chains onto the porous surface, thereby improving the material's affinity for CO_2 [57].

Impregnation: This process also involves several steps to obtain the desired product. It begins with dissolving the chosen precursor of the active metal oxide in a suitable solvent. The porous support material is then introduced into this solution, initiating the impregnation process. During this step, it is important to ensure that the solution penetrates deeply into the pores of the support to achieve effective loading. After impregnation, the material is subjected to calcination, which removes the solvent and leads to the formation of finely dispersed nanoparticles uniformly distributed across the surface of the substrate [58].

Coprecipitation: This approach is well known for producing finely structured, highly interwoven oxide-based multicomponent materials. In this method, different metal salts, selected according to the desired product, are dissolved together in a suitable solvent. Once the precursors are fully dissolved, a precipitating agent such as sodium hydroxide is added, which drives the precipitation of metal species and leads to the formation of hydroxides or carbonates. The resulting product is then thoroughly washed, dried, and finally subjected to calcination, yielding a metal-oxide nanocomposite [59]. These nanocomposites typically exhibit a high degree of dispersion at both the atomic and nanoscale levels, making them highly effective for various applications.

The essence of the synthesis approaches discussed above goes beyond simply following sequential steps. These methods embody carefully chosen strategies that enable the development of highly efficient nanocomposites. Achieving future success in this field will depend on advancing these techniques to deliver precise control while ensuring scalability, industrial feasibility, and environmental sustainability. Such progress will be key to bridging the gap between laboratory-scale methods and large-scale industrial production.

4.5.3 Thermo-mechanical and Morphological Properties of Nanocomposites

The structural identity of nanocomposites is closely tied to their morphology. It is through morphology that we gain insights into the size, shape, concentration, spatial distribution, and orientation of the nanophase within the host matrix. The phase and structure established during synthesis act as the blueprint of the material, ultimately governing both its physical and chemical properties. Equally important is the interfacial area between the fillers and the matrix, which plays a decisive role in defining composite behaviour at the nanoscale. The interface is the region within the polymer matrix where the presence of nanofillers dominates, dictating how the material responds to thermal, mechanical, and environmental stimuli [60]. This underscores the importance of thoroughly understanding morphology as the critical link connecting synthesis, structure, and performance. Such knowledge enables the transition from empirical trial-and-error to rational design, paving the way for predictive modelling and the creation of materials with optimised performance for CCUS applications.

4.5.4 Core Pillars of Nanocomposite Morphology

Nanofiller Geometry: The Impact of Size and Aspect Ratio: Nanocomposites are defined by the incorporation of nanofillers, which must have at least one dimension within the nanoscale regime. This is essential because the so-called "nano-effect" in the matrix can only emerge when the fillers themselves lie in the nanometre range. The size of these fillers, and consequently their volume, directly influences the resulting properties due to their strong interdependence. As the particle size decreases, the surface-to-volume ratio increases dramatically. For spherical particles, this ratio follows an inverse relationship with radius (3/r). To illustrate, a particle with a radius of 10 nm has a specific surface area that is 100 times larger than that of a particle with a radius of 1 µm, assuming both are made of the same material. This sharp increase in surface area at the nanoscale ensures much closer interaction between the fillers and the polymer matrix, thereby generating significant interfacial regions. Beyond just size, the aspect ratio of the nanofillers plays a critical role in determining mechanical reinforcement as well as transport-related properties of the polymer matrix [61]. Among these three classes, 1D and 2D nanofillers, with their high-aspect ratios, generally exhibit superior mechanical reinforcement compared to 0D fillers. Their advantage lies in their ability to span large distances within the polymer matrix, acting like nanoscale bridges that resist crack propagation. For example, CNTs are both extremely strong and stiff. Because of their large length relative to diameter, even a very small fraction of CNTs can form a network inside the polymer. This network effectively transfers stress across microcracks, preventing the matrix from failing prematurely [62]. As a result, remarkable improvements in stiffness and strength can be observed with nanofiller loadings of less than 1 wt%. This remarkable improvement at such low concentrations is widely known as the "nano-reinforcement effect."

4.5.5 The State of Dispersion: The Perennial Challenge of Agglomeration

In the context of nanocomposites, dispersion refers to how well the nanofillers are distributed throughout the host matrix or, in simpler terms, the ability to break apart agglomerated clusters and spread the fillers evenly. Achieving a uniform dispersion has always been one of the central goals of synthesis strategies, as it directly determines the performance of the final material. Despite this focus, attaining proper dispersion remains a major challenge, both scientifically and from an engineering standpoint. The difficulty arises because nanoparticles, with their extremely high surface area, naturally tend to clump together into low-energy clusters rather than staying individually distributed in the matrix [47]. Overcoming this tendency requires strong driving forces and carefully designed methods. Depending on the level of distribution achieved, different types of dispersion can be defined, each linked to a particular structure and performance spectrum. These categories are briefly outlined as follows:

Agglomerated: In this state, the nanofillers tend to cluster together, forming large aggregates. Because of this clustering, the polymer matrix cannot penetrate between individual nanofillers, which drastically reduces the interfacial area. As a result, little to no enhancement in material properties is observed. In fact, these agglomerated clusters often act as weak spots that initiate cracks within the matrix, leading to a decline in key properties such as toughness, tensile strength, and durability.

Intercalated: This morphology is commonly seen with layered fillers such as clays or graphene. Here, instead of dispersing individual nanofillers, the polymer chains penetrate between the layers of the filler. The chains occupy the van der Waals gaps, pushing the layers apart by increasing gallery spacing, while maintaining their stacked arrangement. Intercalated structures represent an intermediate stage of dispersion: more surface area of the filler is exposed compared to the agglomerated state, leading to noticeable improvements in material properties.

Exfoliated: This is considered the ideal state of dispersion. In exfoliated nanocomposites, the fillers are completely separated at the nanoscale and evenly distributed throughout the polymer matrix, without any clustering. This maximises the interfacial area, allowing efficient stress transfer and creating tortuous diffusion pathways that improve barrier properties. At this stage, the thermal, mechanical, and chemical performance of the material reaches its highest potential. Beyond dispersion and orientation, the alignment of nanofillers plays a critical role in tuning material properties. Nanocomposites prepared via solution casting often contain randomly oriented fillers, producing isotropic materials with uniform properties in all directions. However, external forces such as high shear or elongation during extrusion and moulding, or through the application of electric or magnetic fields, can align fillers along a specific direction. Such alignment dramatically enhances properties like stiffness, strength, thermal conductivity, and electrical conductivity along the alignment axis. This directional control is especially valuable for engineering high-performance components tailored for specialised applications [48].

4.5.6 Thermo-Mechanical Properties of Nanocomposites: Performance Under Load and Heat

Their thermo-mechanical properties describe the way nanocomposites respond to heat and mechanical forces. These properties are strongly influenced by the type, size, and distribution of nanofillers within the material. When nanofillers are incorporated into a composite, they can reinforce the structure and dissipate energy at the nanoscale, leading to significant changes in overall performance. Understanding these thermo-mechanical properties is crucial because they determine not only how a material will behave under different conditions but also how long it will last in service [63, 64]. A clear grasp of these factors allows researchers and engineers to predict performance and optimise nanocomposites for targeted applications. Some of the most important thermo-mechanical properties of nanocomposites are outlined below:

Thermal Stability and Degradation Behaviour: Thermal stability refers to a material's ability to retain its structural integrity without deforming or breaking down at elevated temperatures. Studies have shown that adding nanofillers can significantly improve the thermal stability of nanocomposites. This property is commonly evaluated using Thermogravimetric Analysis (TGA), where the weight loss of a material is measured as the temperature increases. Compared to pure polymers, nanocomposites typically show a higher onset temperature of degradation and a slower rate of weight loss. These improvements arise from a combination of physical and chemical mechanisms, which can be illustrated as follows:

Barrier Mechanism: When nanofillers with a high-aspect ratio, such as graphene or clays, are uniformly dispersed in the polymer matrix, they create tortuous and winding pathways inside the material. These meandering paths act like obstacles, restricting the movement of small volatile molecules trying to escape from the bulk of the material. At the same time, they also slow down the diffusion of atmospheric oxygen into the polymer. Together, these effects reduce the rate of thermal-oxidative degradation, thereby improving the material's durability [65].

Enhanced Heat Dissipation: Nanofillers with inherently high thermal conductivity, once incorporated into the polymer, greatly enhance the overall thermal conductivity of the composite. This allows heat to be distributed more evenly throughout the material, preventing localised hotspots. Without these fillers, heat tends to accumulate in one place, often triggering early degradation. By spreading heat uniformly, the nanofillers help the composite withstand higher thermal stress [66].

Catalytic Char Formation: High surface area nanofillers also act as catalytic sites within the nanocomposite. They redirect the usual degradation pathways by promoting condensation and cross-linking reactions, which lead to the formation of a thermally stable carbonaceous char layer. This char serves as a physical barrier, reducing mass transport and acting as an insulating shield that protects the underlying polymer layers from further heat exposure. The flame-retardant properties often observed in nanocomposites primarily arise from this mechanism [65].

4.5.7 Glass Transition Temperature

Glass transition (T_g) is one of the most fundamental transformations in amorphous polymer matrices. At this temperature, the polymer changes from a rigid, glassy state to a softer, more flexible state. As the material approaches T_g, the long polymer chains begin to move cooperatively. When nanofillers are introduced into the polymer matrix, the situation becomes more interesting. At the interfaces, where polymer chains interact with the nanofillers, T_g is often higher because of the strong interactions at the filler surface [66]. To study T_g, researchers commonly use Differential Scanning Calorimetry (DSC) or Dynamic Mechanical Analysis (DMA). The changes observed in the T_g curves reveal important insights:

1. If the T_g shifts upward (increases), it means there are strong attractive interactions between polymer chains and the nanofillers. These interactions physically restrict the mobility of the chains at the interface. As a result, more thermal energy is required to initiate motion, leading to a higher transition temperature.

2. If T_g shifts downward (decreases), it suggests that the interaction between the filler and polymer is weak or even repulsive. In this case, the polymer chains are less tightly packed, creating additional free volume. This looser packing allows chains to move more easily, resulting in a lower transition temperature.

3. The shape of the T_g transition in DSC or DMA curves also carries meaning. A broader transition indicates a heterogeneous environment within the polymer, where chain mobility varies gradually, from strongly restricted near the filler interface to free in the bulk. In contrast, a sharper transition point to a more uniform environment with less variation in chain mobility.

4.5.8 Thermal Conductivity

Polymers, by nature, are usually poor conductors of heat and are considered thermal insulators. However, with the development of nanocomposites, it has become possible to improve their thermal properties significantly. The key factor enabling this improvement is the incorporation of nanofillers. The thermal conductivity of polymer nanocomposites is achieved when these nanofillers form a percolating network within the polymer matrix. At low concentrations, the fillers remain isolated and separated by the insulating polymer, which prevents efficient heat flow. But as the filler concentration increases and reaches a critical threshold (known as the percolation limit), the fillers connect with each other, forming a continuous pathway. This network allows the smooth transfer of phonons, the primary carriers of heat in solids, thereby making the nanocomposite significantly more thermally conductive. Interestingly, the shape of the fillers plays a major role. Fillers with a high-aspect ratio (like CNTs or graphene sheets) reach this percolation threshold at much lower concentrations compared to spherical fillers. This is because thin, elongated, or sheet-like fillers can form interconnected networks more efficiently, even at small volume fractions [67].

Mechanical Performance and Reinforcement: A key priority in developing structural nanocomposites is to enhance their mechanical properties. Strong mechanical performance is what allows these materials to function effectively in load-bearing environments. The overall strength and durability of a nanocomposite largely depend on several critical factors that govern how well the reinforcement works [68, 69]. Some of the main factors are briefly outlined below:

Young's Modulus: Young's Modulus is the key factor that determines the stiffness of a material. It is defined as the ability of a material to undergo elastic deformation under applied stress. When comparing polymers and nanofillers, nanofillers are usually much stiffer. For example, the Young's modulus of epoxy is about 3 GPa, whereas that of CNTs is nearly 300 times higher. When nanofillers are incorporated into the polymer matrix, they significantly enhance the overall stiffness (modulus) of the resulting composite. However, the final modulus of the nanocomposite does not depend on filler stiffness alone. It also depends on the filler's volume fraction, aspect ratio, and, most importantly, how effectively stress is transferred across the filler-matrix interface. This relationship is a well-established concept in nanocomposite mechanics and is often analysed or predicted using models such as the Halpin-Tsai model.

Tensile Strength: Tensile strength is defined as the maximum stress a material can withstand before it undergoes structural distortion or fracture. Unlike stiffness, the effect of nanofillers on strength is more complex and highly sensitive to the overall morphology of the nanocomposite. The improvement in strength largely depends on how efficiently stress is transferred from the polymer matrix to the nanofillers. This requires two critical factors: uniform dispersion of the fillers and strong interfacial adhesion between the filler and matrix. If the fillers agglomerate, they act like built-in defects within the material, creating points where stress becomes concentrated. These stress concentration sites can trigger cracks or fractures even under relatively low loads, thereby reducing the overall strength. Similarly, if the interfacial adhesion between the polymer and filler is weak, the bond fails to transfer stress effectively. In such cases, even small loads can cause premature failure of the composite. For these reasons, achieving high strength in nanocomposites is far more challenging than improving stiffness. A nanocomposite that exhibits both high strength and uniform structure is regarded as an optimally synthesised material, reflecting the best balance of processing and material design [69].

Toughness, Ductility and Fracture Behaviour: The three mentioned parameters are closely related. Toughness is an energy value which a material can absorb without being damaged and is estimated by evaluating the area under the stress-strain curve. Ductility is the capability of a material to which it can be stretched without any damage. In general, it has been observed that toughness and the strength of the materials are inversely related to each other; that is, stronger materials often prove to be brittle in nature. However, in the case of the nanocomposite, this is balanced through activation of different energy dissipation mechanisms at the nano-regime. This activation results in a simultaneous increase of both parameters [70]. Some of the prominent mechanisms involved in toughness are briefly explained as:

Crack Bridging: Nanofillers with a high-aspect ratio can act like tiny bridges across a crack, holding the two separated faces together and preventing the crack from spreading further. For the crack to progress, it would either have to break through the filler or force it out of the matrix; both processes require a lot of energy, which makes crack propagation very difficult.

Crack Tip Blunting and Deflection: When a growing crack encounters a nanoparticle, it faces resistance. This interaction generates local stress, which is relieved by forcing the crack to change its direction. As a result, the crack follows a more zigzag path, consuming more energy and making the fracture tougher to complete.

Matrix Crazing and Shear Deflection: In the regions near the nanoparticles, the stress field from a crack can cause the polymer matrix to undergo plastic deformation, such as forming micro-voids (crazes) or shear bands. These deformations absorb a large amount of energy, further slowing down crack growth.

Interfacial Debonding and Filler Pull Out: In composites with optimised filler-matrix bonding, cracks may cause partial debonding at the interface. When this happens, the fillers slide or pull out from the matrix. This sliding not only resists fracture but also dissipates energy as heat, further enhancing toughness.

4.5.9 Dynamic Mechanical Analysis (DMA)

As mentioned above, DMA is one of the prominent tools or techniques for analysing the thermo-mechanical behaviour of nanocomposites. This mechanism elucidates the viscoelastic nature of the material through the application of oscillating stress. The effect of the applied stress is measured in the generated strain as a function of temperature [66]. The analysis of DMA provides us with information about the following parameters:

Storage Modulus (E′): This represents how stiff or elastic a material is. When we look at a DMA plot of E′ versus temperature, it gives us a complete picture of how stiffness changes as the material is heated. The curve typically shows three main regions: a stiff, high-modulus "glassy" plateau at low temperatures, a sharp drop around the glass transition, and finally a softer, low-modulus plateau at higher temperatures. The presence of nanofillers usually shows up as a noticeable increase in E′, meaning the material stays stiffer across a wider temperature range.

Loss Modulus (E″): This measures how much energy is lost as heat during each cycle of deformation. On the DMA curve, the peak of E″ is strongly linked to the glass transition, since this is where molecular motion in the polymer absorbs the most energy.

Damping Factor (tan δ): The ratio of E″ to E′ (tan δ) tells us about the material's damping ability, how well it balances energy storage and dissipation. The peak of the tan δ curve corresponds to the glass transition temperature. The height and width of this peak also provide valuable insights: a taller, sharper peak indicates strong interfacial constraints and uniform dynamics, while a broader peak points to heterogeneity and a gradient of mobility within the polymer. From the discussion of the thermo-mechanical and morphological properties of nanocomposite materials, it is clear that these two aspects are deeply interconnected in a hierarchical way. The physics of the synthesis process shapes the morphology, while the chemistry of that morphology ultimately governs the structure of the material. This structure then acts as the foundation, dictating key thermo-mechanical behaviours such as heat conduction, stress transfer, and energy dissipation. To truly understand these complex interconnections, a holistic approach is essential. Only by exploring the full structure-property relationship can we design nanocomposites more intelligently, paving the way for sustainable, high-performance materials that can meet the demands of future engineering applications.

4.5.10 Limitations

The rapid progress and development of nanocomposites have emerged as a frontier in materials science. The excellent properties of these materials have revolutionised carbon-capturing technology. However, their transition from the laboratory-scale to the industrial sector is facing a lot of challenges and inherent limitations. Also, the properties that require control at the nanoscale regime have posed many complexities in their fabrication. Besides this, controlling this regime has critical effects on its longevity, cost-effectiveness, and eco-friendliness. Thus, it is necessary to have a clear understanding of these aspects, which will be beneficial for carrying out future research aiming towards practical and sustainable solutions.

4.5.11 Challenges in Synthesis and Scalability

The first and most crucial step in designing nanocomposites is their synthesis. Achieving desirable properties and well-defined morphology requires precise control over the synthesis parameters. Only with such precision can a uniform nanofiller distribution within the nanocomposite structure be obtained. Some of the most prominent and persistent challenges include:

1. A common tendency in nanomaterials is the minimisation of surface energy to achieve stability. Nanoparticles, with their inherently high surface area, often agglomerate when incorporated into a polymer matrix as a means of reducing exposed surface area. Thermodynamic forces drive this process. Overcoming these forces to achieve a uniform distribution of nanofillers requires significant energy input and often additional steps, such as the use of surfactants or chemical modifications. While these strategies can improve dispersion, they complicate the synthesis process, increase energy consumption, and raise production costs. At the industrial scale, such approaches are even less efficient and cost-effective. Moreover, the presence of mini-agglomerates in large-scale composites can act as stress concentration points, weakening the mechanical integrity of the material. In membrane-based applications, these agglomerates can also form non-selective voids, severely reducing separation efficiency.

2. Another challenge arises from the synthesis of nanofillers commonly used for carbon capture, such as CNTs, graphene nanosheets, and MOFs. Their preparation is typically a multistep, energy-intensive process, which drives up production costs. As a result, the incorporation of these expensive nanofillers into nanocomposites makes the final material costly, limiting its large-scale use in applications like carbon capture that require substantial material quantities.

3. Scalability remains a critical issue. Many synthesis routes that yield high-performance nanocomposites at the laboratory scale prove difficult to reproduce efficiently on an industrial scale. Furthermore, maintaining consistent material characteristics across different production batches is challenging. Even small variations in synthesis parameters such as temperature, pressure, concentration, or mixing time can lead to noticeable differences in the final product. This batch-to-batch variability presents a significant barrier to commercialisation, where uniformity and reproducibility are essential.

4.5.12 Long-Term Stability and Durability

After the synthesis of nanocomposites, the stability and durability can vary from application to application, as different applications are carried out under different environmental conditions. In view of employing the materials for carbon capture, their stability is a major concern, as this process is being carried out under high pressure and temperature, and even in the presence of corrosive acidic gases. Some of the prominent concerns that need to be taken care of are briefly explained as follows:

1. Although nanocomposites generally exhibit improved thermal stability compared to pristine polymers, they remain susceptible to degradation under prolonged exposure to elevated temperatures. The stability issues vary depending on the type of nanocomposite. For instance, in some systems, functional groups may lose activity or detach at high temperatures. MOFs are particularly vulnerable to moisture and acidic environments, where the bonds between metal nodes and organic linkers can break, leading to structural collapse. Similarly, biopolymer-based nanocomposites often degrade at temperatures comparable to those encountered in flue gases, limiting their long-term applicability under such conditions.

2. For membrane-based nanocomposites, a major challenge lies in maintaining mechanical integrity over repeated operational cycles and extended periods of use. As discussed earlier, the incorporation of nanofillers can sometimes increase stiffness, which in turn reduces flexibility and overall durability. Additionally, physical ageing, arising from the gradual rearrangement of polymer chains, leads to a reduction in free volume, thereby lowering permeability. In synthesis approaches that involve pressure swings, delamination at the filler-matrix interface may occur, compromising both selectivity and membrane lifespan. Similarly, in adsorbent-based nanocomposites, issues such as physical attrition of pellets and powders, or fluidisation in bed reactors, progressively degrade performance over time.

3. Another critical challenge for carbon-capturing nanocomposite materials is ensuring reusability over a large number of cycles with only minimal loss in efficiency, which is essential for maintaining economic viability. While many materials demonstrate excellent initial performance, their effectiveness often declines during repeated adsorption–desorption cycles. The ability to retain capacity and stability over extended use, commonly referred to as long-term cyclability, is therefore vital for CCUS applications. This necessitates the development of robust nanocomposites supported by optimised regeneration protocols to ensure sustained operational performance.

Performance in Realistic Conditions: So far, most of the materials reported are being tested in a laboratory, where they are employed under suitable atmospheric conditions. However, in reality, the scenario is altogether different for real-world applications, which pose challenges such as:

1. Industrial flue gases, from which CO_2 must be captured, are typically complex mixtures containing substances such as NO_2, water vapour, oxygen, and acidic trace components. This complexity significantly reduces the efficiency of nanocomposite-based capture materials. For example, water vapour competes with CO_2 during adsorption, particularly in MOFs and amine-functionalised materials, leading to reduced capture performance. Similarly, acidic impurities can irreversibly react with basic active sites, permanently deactivating them and suppressing adsorption capacity. While pre-removal of these impurities is possible, it requires additional processing steps that consume energy and increase both the cost and complexity of the overall capture process.

2. Another critical challenge is the inherent trade-off between adsorption capacity and selectivity. Although nanocomposites improve fundamental material properties, optimisation often comes at the expense of one characteristic over another. In membranes, for instance, high permeability usually results in reduced selectivity, making it difficult to achieve both simultaneously. Similarly, adsorbents with strong CO_2 binding sites exhibit excellent uptake but require high regeneration energy, thereby raising operational costs. Achieving a balance that ensures both efficiency and economic feasibility remains one of the key challenges in the development of carbon capture nanocomposites.

Environmental, Health, and Safety (EHS) Concerns: Like any class of materials involving chemical entities, nanocomposites come with both advantages and limitations. A critical concern is their potential impact on health and the environment, which necessitates careful safety assessments. One of the foremost challenges is the toxicity associated with nanoparticles. Prolonged inhalation of nanoparticles has been shown to cause chronic health effects. In addition, during processing, some nanomaterials may leach into wastewater streams, ultimately entering local water bodies and posing risks to both aquatic ecosystems and human health. To mitigate these risks, appropriate engineering controls and workplace safety protocols must be established. Beyond occupational safety, comprehensive life-cycle analyses are essential to evaluate the broader environmental impact of nanocomposites. Factors such as energy consumption during synthesis, the type and availability of precursors, end-of-life disposal, and recycling strategies must be carefully considered. Without such evaluations, there is a real possibility that in solving one problem, such as CO_2 capture, we may inadvertently create new and equally serious environmental or health concerns from the overall analysis. Nanocomposites hold great promise for the future of carbon capture technologies. However, significant progress and advancements are still required to realise their full potential. Key challenges that must be addressed include developing cost-effective synthesis methods, ensuring long-term stability, and designing environmentally sustainable processes. A robust and holistic approach that integrates efficiency, affordability, and eco-friendliness is therefore essential to establish nanocomposites as viable materials for a sustainable future.

4.6 Conclusion

Addressing the pressing challenge of climate change requires innovative technologies, and the development of nanocomposite-based carbon capture materials has already made a significant impact. This chapter has provided a comprehensive overview of the different materials and methods explored to date for this purpose. Particular emphasis has been placed on the synergistic combination of nanofillers with processable matrices, highlighting how such integrations have led to remarkable improvements in performance. The discussion also underscores the importance of carefully controlling synthesis parameters, as precise regulation of morphology directly influences the thermo-mechanical stability of the materials and, consequently, their CO_2 capturing efficiency. Nevertheless, several challenges remain, particularly in translating laboratory-scale successes to industrial applications. Future progress will require a meticulous, holistic, and pragmatic approach that not only demonstrates proof-of-concept in the laboratory but also ensures scalability, cost-effectiveness, and environmental sustainability. Only through such a vision can nanocomposite-based carbon capture technologies move from promising research to practical solutions for mitigating climate change.

REFERENCES

[1] B. Ganesh Guptha, K. Koteswara Rao, Nano Composites for Carbon Capture-A Brief Review, Int. J. Res. Publ. Rev. (2025). www.ijrpr.com.

[2] T.A. Saleh, Nanomaterials and hybrid nanocomposites for CO_2 capture and utilization: environmental and energy sustainability, RSC Adv. 12 (2022) 23869. https://doi.org/10.1039/D2RA03242B.

[3] R.M. Cigala, G. De Luca, I. Ielo, F. Crea, Biopolymeric Nanocomposites for CO_2 Capture, Polymers (Basel). 16 (2024). https://doi.org/10.3390/POLYM16081063.

[4] F.A. Ibekwe, H.S. Samuel, D.A. Undie, O.O. Akinpelu, O.P. Onotu, E.E. Etim, The nanogreen revolution: Transforming CO_2 capture through sustainable nanotechnology, Next Energy 9 (2025) 100395. https://doi.org/10.1016/J.NXENER.2025.100395.

[5] K.M.G. Langie, K. Tak, C. Kim, H.W. Lee, K. Park, D. Kim, W. Jung, C.W. Lee, H.S. Oh, D.K. Lee, J.H. Koh, B.K. Min, D.H. Won, U. Lee, Toward economical application of carbon capture and utilization technology with near-zero carbon emission, Nat. Commun. 13 (2022). https://doi.org/10.1038/S41467-022-35239-9.

[6] Circular Carbon Economy and sustainability | Aramco. https://www.aramco.com/en/sustainability/climate-and-energy/circular-carbon-economy?utm.

[7] M.M. Sadiq, K. Konstas, P. Falcaro, A.J. Hill, K. Suzuki, M.R. Hill, Engineered Porous Nanocomposites That Deliver Remarkably Low Carbon Capture Energy Costs, Cell Reports Phys. Sci. 1 (2020) 100070. https://doi.org/10.1016/J.XCRP.2020.100070.

[8] Y.J. Park, H. Lee, H.L. Choi, M.C. Tapia, C.Y. Chuah, T.H. Bae, Mixed-dimensional nanocomposites based on 2D materials for hydrogen storage and CO_2 capture, Npj 2D Mater. Appl. 7 (2023) 1–18. https://doi.org/10.1038/S41699-023-00425-W;SUBJMETA.

[9] V. Ramar, A. Balraj, Critical Review on Carbon-Based Nanomaterial for Carbon Capture: Technical Challenges, Opportunities, and Future Perspectives, Energy & Fuels 36 (2022) 13479–13505. https://doi.org/10.1021/ACS.ENERGYFUELS.2C02585.

[10] X. Hu, H. Cong, Y. Shen, M. Radosz, Nanocomposite Membranes for CO_2 Separations: Silica/Brominated Poly(phenylene oxide), Ind. Eng. Chem. Res. 46 (2007) 1547–1551. https://doi.org/10.1021/IE061055Y.

[11] A.A. Alhwaige, H. Ishida, S. Qutubuddin, Carbon Aerogels with Excellent CO_2 Adsorption Capacity Synthesized from Clay-Reinforced Biobased Chitosan-Polybenzoxazine Nanocomposites, ACS Sustain. Chem. Eng. 4 (2016) 1286–1295. https://doi.org/10.1021/ACSSUSCHEMENG.5B01323.

[12] S. Castro-Pardo, S. Bhattacharyya, R.M. Yadav, A.P. de Carvalho Teixeira, M.A. Campos Mata, T. Prasankumar, M.A. Kabbani, M.G. Kibria, T. Xu, S. Roy, P.M. Ajayan, A comprehensive overview of carbon dioxide capture: From materials, methods to industrial status, Mater. Today 60 (2022) 227–270. https://doi.org/10.1016/J.MATTOD.2022.08.018.

[13] M.S. Al-Bukhari, I. Abdulazeez, M.M. Abdelnaby, I.H. Aljundi, O.C.S. Al Hamouz, 3D porous polymers for selective removal of CO_2 and H_2 storage: experimental and computational studies, Front. Chem. 11 (2023) 1265324. https://doi.org/10.3389/FCHEM.2023.1265324/BIBTEX.

[14] J. Jin, Z. Wen, S. Li, J. Huang, Quantum chemical study of CO_2 physisorption and chemisorption on EDA-grafted graphene oxide, Greenh. Gases Sci. Technol. 13 (2023) 357–368. https://doi.org/10.1002/GHG.2201.

[15] B. Dziejarski, J. Serafin, K. Andersson, R. Krzyżyńska, CO_2 capture materials: a review of current trends and future challenges, Mater. Today Sustain. 24 (2023) 100483. https://doi.org/10.1016/J.MTSUST.2023.100483.

[16] R. Ma, Y. Hou, W. Zhang, Y.T. Pan, S. Huo, · Congling Shi, MOF-based nanocomposites in polymer matrix: progress and prospects, Adv. Compos. Hybrid Mater. 2025 85 8 (2025) 1–30. https://doi.org/10.1007/S42114-025-01435-6.

[17] M.G. Kotp, I.M. Minisy, B. Al-Saida, S.W. Kuo, Kinetics and selectivity insights into carbon dioxide capture utilizing carboxymethyl cellulose-polypyrrole nanocomposites: Screening of silane functionalization, Carbohydr. Polym. 356 (2025) 123399. https://doi.org/10.1016/J.CARBPOL.2025.123399.

[18] T. Khandaker, M.S. Hossain, P.K. Dhar, M.S. Rahman, M.A. Hossain, M.B. Ahmed, Efficacies of Carbon-Based Adsorbents for Carbon Dioxide Capture, Process. 2020, Vol. 8, Page 654 8 (2020) 654. https://doi.org/10.3390/PR8060654.

[19] D. Kichukova, T. Lazarova, G. Atanasova, D. Kovacheva, I. Spassova, Tailored Carbon Nanocomposites for Efficient CO2 Capture, Molecules 30 (2025) 2408. https://doi.org/10.3390/MOLECULES30112408/S1.

[20] D.M. Glenna, A. Jana, Q. Xu, Y. Wang, Y. Meng, Y. Yang, M. Neupane, L. Wang, H. Zhao, J. Qian, S.W. Snyder, Carbon Capture: Theoretical Guidelines for Activated Carbon-Based CO2 Adsorption Material Evaluation, J. Phys. Chem. Lett. 14 (2023) 10693–10699. https://doi.org/10.1021/ACS.JPCLETT.3C02711.

[21] X. Li, Y. Cheng, H. Zhang, S. Wang, Z. Jiang, R. Guo, H. Wu, Efficient CO_2 Capture by Functionalized Graphene Oxide Nanosheets as Fillers To Fabricate Multi-Permselective Mixed Matrix Membranes, ACS Appl. Mater. Interfaces 7 (2015) 5528–5537. https://doi.org/10.1021/ACSAMI.5B00106.

[22] J. Hack, N. Maeda, D.M. Meier, Review on CO_2 Capture Using Amine-Functionalized Materials, ACS Omega 7 (2022) 39520–39530. https://doi.org/10.1021/ACSOMEGA.2C03385.

[23] I.A.W. Tan, A.L. Ahmad, B.H. Hameed, Adsorption of basic dye on high-surface-area activated carbon prepared from coconut husk: Equilibrium, kinetic and thermodynamic studies, J. Hazard. Mater. 154 (2008) 337–346. https://doi.org/10.1016/J.JHAZMAT.2007.10.031.

[24] Y. Wang, T.G. Feric, J. Tang, C. Fang, S.T. Hamilton, D.M. Halat, B. Wu, H. Celik, G. Rim, T. DuBridge, J. Oshiro, R. Wang, A.H.A. Park, J.A. Reimer, Carbon capture in polymer-based electrolytes, Sci. Adv. 10 (2024). https://doi.org/10.1126/sciadv.adk2350.

[25] R. Lin, L. Ge, L. Hou, E. Strounina, V. Rudolph, Z. Zhu, Mixed Matrix Membranes with Strengthened MOFs/Polymer Interfacial Interaction and Improved Membrane Performance, ACS Appl. Mater. Interfaces 6 (2014) 5609–5618. https://doi.org/10.1021/AM500081E.

[26] S. Mahajan, M. Lahtinen, Recent progress in metal-organic frameworks (MOFs) for CO_2 capture at different pressures, J. Environ. Chem. Eng. 10 (2022) 108930. https://doi.org/10.1016/J.JECE.2022.108930.

[27] A.F.; Yazid, H.; Mukhtar, R.; Nasir, D.F. Mohshim, Z. Xie, A. Farzana Yazid, H. Mukhtar, R. Nasir, D.F. Mohshim, Incorporating Carbon Nanotubes in Nanocomposite Mixed-Matrix Membranes for Gas Separation: A Review, Membr. 2022, Vol. 12, Page 589 12 (2022) 589. https://doi.org/10.3390/MEMBRANES12060589.

[28] T. Xia, Y. Wu, T. Ji, W. Hu, K. Yu, X. He, B.H. Yin, Y. Liu, Mixed-matrix membranes incorporating hierarchical ZIF-8 towards enhanced CO_2/N_2 separation, Smart Mol. 3 (2025) e20240066. https://doi.org/10.1002/smo.20240066.

[29] K.C. Wong, P.S. Goh, A.F. Ismail, H.S. Kang, Q. Guo, X. Jiang, J. Ma, The State-of-the-Art Functionalized Nanomaterials for Carbon Dioxide Separation Membrane, Membr. 2022, Vol. 12, Page 186 12 (2022) 186. https://doi.org/10.3390/MEMBRANES12020186.

[30] A. Banerjee, D. Das, R. Andler, R. Bandopadhyay, Green Synthesis of Silver Nanoparticles Using Exopolysaccharides Produced by Bacillus anthracis $PFAB_2$ and Its Biocidal Property, J. Polym. Environ. 29 (2021) 2701–2709. https://doi.org/10.1007/S10924-021-02051-3/TABLES/1.

[31] G. Yang, H. Luo, T. Ohba, H. Kanoh, CO_2 Capture by Carbon Aerogel–Potassium Carbonate Nanocomposites, Int. J. Chem. Eng. 2016 (2016) 4012967. https://doi.org/10.1155/2016/4012967.

[32] Y. Chen, Y. Zhao, J. Zhang, C. Zheng, Hydrogen production through CO_2 sorption-enhanced methane steam reforming: Comparison between different adsorbents, Sci. China Technol. Sci. 54 (2011) 2999–3008. https://doi.org/10.1007/S11431-011-4587-6/METRICS.

[33] T.M. Nyathi, N. Fischer, A.P.E. York, D.J. Morgan, G.J. Hutchings, E.K. Gibson, P.P. Wells, C.R.A. Catlow, M. Claeys, Impact of Nanoparticle–Support Interactions in Co_3O_4/Al_2O_3 Catalysts for the Preferential Oxidation of Carbon Monoxide, ACS Catal. 9 (2019) 7166–7178. https://doi.org/10.1021/ACSCATAL.9B00685.

[34] P. Murge, S. Dinda, S. Roy, Zeolite-Based Sorbent for CO_2 Capture: Preparation and Performance Evaluation, Langmuir (2019). https://doi.org/10.1021/ACS.LANGMUIR.9B02259.

[35] J.T. Anyanwu, Y. Wang, R.T. Yang, SBA-15 Functionalized with Amines in the Presence of Water: Applications to CO_2 Capture and Natural Gas Desulfurization, Ind. Eng. Chem. Res. 60 (2021) 6277–6286. https://doi.org/10.1021/ACS.IECR.1C00415.

[36] J.R. Li, R.J. Kuppler, H.C. Zhou, Selective gas adsorption and separation in metal-organic frameworks, Chem. Soc. Rev. 38 (2009) 1477–1504. https://doi.org/10.1039/b802426j.

[37] Q. Wang, J. Luo, Z. Zhong, A. Borgna, CO_2 capture by solid adsorbents and their applications: current status and new trends, Energy Environ. Sci. 4 (2010) 42–55. https://doi.org/10.1039/C0EE00064G.

[38] D.M. D'Alessandro, B. Smit, J.R. Long, Carbon dioxide capture: Prospects for new materials, Angew. Chemie - Int. Ed. 49 (2010) 6058–6082. https://doi.org/10.1002/anie.201000431.

[39] A.M. Marti, S.R. Venna, E.A. Roth, J.T. Culp, D.P. Hopkinson, Simple Fabrication Method for Mixed Matrix Membranes with in Situ MOF Growth for Gas Separation, ACS Appl. Mater. Interfaces 10 (2018) 24784–24790. https://doi.org/10.1021/ACSAMI.8B06592.

[40] X. Jia, Z. Song, Q. Li, J. Huang, X. Zhai, L. Tian, J. Li, Z. Qiao, Y. Luo, D. Zhang, One-pot synthesis of necklace-like MOF@CNTs: a universal strategy for enhancing molecular separation performance in mixed-matrix membranes, RSC Appl. Interfaces 2 (2025) 1041–1050. https://doi.org/10.1039/d5lf00016e.

[41] A.A. Tourzani, F. Hormozi, M. Asadollahzadeh, R. Torkaman, Effective CO_2 capture by using poly (acrylonitrile) nanofibers based on the radiation grafting procedure in fixed-bed adsorption column, Sci. Rep. 13 (2023) 1–15. https://doi.org/10.1038/S41598-023-33036-Y;SUBJMETA.

[42] Y.M. Manawi, Ihsanullah, A. Samara, T. Al-Ansari, M.A. Atieh, A Review of Carbon Nanomaterials' Synthesis via the Chemical Vapor Deposition (CVD) Method, Mater. 2018, Vol. 11, Page 822 11 (2018) 822. https://doi.org/10.3390/MA11050822.

[43] X. Hu, M. Gholizadeh, Biomass pyrolysis: A review of the process development and challenges from initial researches up to the commercialisation stage, J. Energy Chem. 39 (2019) 109–143. https://doi.org/10.1016/J.JECHEM.2019.01.024.

[44] M. Sevilla, A.B. Fuertes, The production of carbon materials by hydrothermal carbonization of cellulose, Carbon N. Y. 47 (2009) 2281–2289. https://doi.org/10.1016/J.CARBON.2009.04.026.

[45] T.K. Junita, N. Syakir, F. Faizal, N. Fitrilawati, Graphene-Based Composite for Carbon Capture, ACS Omega 9 (2024) 20658–20669. https://doi.org/10.1021/ACSOMEGA.3C08722.

[46] Y.T. Shieh, T.T. Hsiao, S.K. Chang, CO_2 pressure effects on melting, crystallization, and morphology of poly(vinylidene fluoride), Polymer (Guildf). 47 (2006) 5929–5937. https://doi.org/10.1016/J.POLYMER.2006.06.022.

[47] M. Alexandre, P. Dubois, Polymer-layered silicate nanocomposites: preparation, properties and uses of a new class of materials, Mater. Sci. Eng. R Reports 28 (2000) 1–63. https://doi.org/10.1016/S0927-796X(00)00012-7.

[48] D.R. Paul, L.M. Robeson, Polymer nanotechnology: Nanocomposites, Polymer (Guildf). 49 (2008) 3187–3204. https://doi.org/10.1016/J.POLYMER.2008.04.017.

[49] L.E. Kreno, K. Leong, O.K. Farha, M. Allendorf, R.P. Van Duyne, J.T. Hupp, Metal–Organic Framework Materials as Chemical Sensors, Chem. Rev. 112 (2011) 1105–1125. https://doi.org/10.1021/CR200324T.

[50] J. Huang, D. Yang, Z. Hu, H. Zhang, Z. Zhang, F. Wang, Y. Xie, S. Liu, Q. Wang, C.U. Pittman, In situ growth of Zn-based metal–organic frameworks in ultra-high surface area nano-wood aerogel for efficient CO2 capture and separation, J. Mater. Chem. A 11 (2023) 16878–16888. https://doi.org/10.1039/D3TA02229C.

[51] J. Cravillon, S. Münzer, S.J. Lohmeier, A. Feldhoff, K. Huber, M. Wiebcke, Rapid Room-Temperature Synthesis and Characterization of Nanocrystals of a Prototypical Zeolitic Imidazolate Framework, Chem. Mater. 21 (2009) 1410–1412. https://doi.org/10.1021/CM900166H.

[52] O. Shekhah, H. Wang, S. Kowarik, F. Schreiber, M. Paulus, M. Tolan, C. Sternemann, F. Evers, D. Zacher, R.A. Fischer, C. Wöll, Step-by-Step Route for the Synthesis of Metal–Organic Frameworks, J. Am. Chem. Soc. 129 (2007) 15118–15119. https://doi.org/10.1021/JA076210U.

[53] D.R. Latulippe, A.M. Mika, R.F. Childs, R. Ghosh, C.D.M. Filipe, Flux performance and macrosolute sieving behavior of environment responsive formed-in-place ultrafiltration membranes, J. Memb. Sci. 342 (2009) 227–235. https://doi.org/10.1016/J.MEMSCI.2009.06.044.

[54] D. Needleman, Z. Dogic, Active matter at the interface between materials science and cell ...biology, Nat. Rev. Mater. 2 (2017) 1–14. https://doi.org/10.1038/NATREVMATS.2017.48;SUBJMETA

[55] D. Zhao, J. Feng, Q. Huo, N. Melosh, G.H. Fredrickson, B.F. Chmelka, G.D. Stucky, Triblock Copolymer Syntheses of Mesoporous Silica with Periodic 50 to 300 Angstrom Pores, Science (80). 279 (1998) 548–552. https://doi.org/10.1126/SCIENCE.279.5350.548.

[56] A.F. Masters, T. Maschmeyer, Zeolites – From curiosity to cornerstone, Microporous Mesoporous Mater. 142 (2011) 423–438. https://doi.org/10.1016/J.MICROMESO.2010.12.026.

[57] H. Zheng, C. Gao, S. Che, Amino and quaternary ammonium group functionalized mesoporous silica: An efficient ion-exchange method to remove anionic surfactant from AMS, Microporous Mesoporous Mater. 116 (2008) 299–307. https://doi.org/10.1016/J.MICROMESO.2008.04.016.

[58] X. Liu, Y. Yang, W. Ji, H. Liu, C. Zhang, B. Xu, Controllable growth of nanostructured carbon from coal tar pitch by chemical vapor deposition, Mater. Chem. Phys. 104 (2007) 320–326. https://doi.org/10.1016/J.MATCHEMPHYS.2007.03.020.

[59] R.O. Hussein, X. Nie, D.O. Northwood, A spectroscopic and microstructural study of oxide coatings produced on a Ti–6Al–4V alloy by plasma electrolytic oxidation, Mater. Chem. Phys. 134 (2012) 484–492. https://doi.org/10.1016/J.MATCHEMPHYS.2012.03.020.

[60] P.M. Ajayan, L.S. Schadler, P. V. Braun, Nanocomposite Science and Technology, Nanocomposite Sci. Technol. (2004) 1–230. https://doi.org/10.1002/3527602127.

[61] E.T. Thostenson, Z. Ren, T.W. Chou, Advances in the science and technology of carbon nanotubes and their composites: a review, Compos. Sci. Technol. 61 (2001) 1899–1912. https://doi.org/10.1016/S0266-3538(01)00094-X.

[62] Z. Spitalsky, D. Tasis, K. Papagelis, C. Galiotis, Carbon nanotube–polymer composites: Chemistry, processing, mechanical and electrical properties, Prog. Polym. Sci. 35 (2010) 357–401. https://doi.org/10.1016/J.PROGPOLYMSCI.2009.09.003.

[63] Z. Latif, M. Ali, E.J. Lee, Z. Zubair, K.H. Lee, Thermal and Mechanical Properties of Nano-Carbon-Reinforced Polymeric Nanocomposites: A Review, J. Compos. Sci. 2023, Vol. 7, Page 441 7 (2023) 441. https://doi.org/10.3390/JCS7100441.

[64] Y.N. Sudhakar, M. Selvakumar, D.K. Bhat, Investigations on thermo-mechanical properties of organically modified polymer clay nanocomposites for packaging application, Polym. Polym. Compos. 29 (2021) 1191–1199. https://doi.org/10.1177/0967391120960645.

[65] C.E. Corcione, M. Frigione, Characterization of Nanocomposites by Thermal Analysis, Materials (Basel). 5 (2012) 2960. https://doi.org/10.3390/MA5122960.

[66] H. Jung, H.K. Choi, Y. Oh, H. Hong, J. Yu, Enhancement of thermo-mechanical stability for nanocomposites containing plasma treated carbon nanotubes with an experimental study and molecular dynamics simulations, Sci. Rep. 10 (2020) 1–10. https://doi.org/10.1038/S41598-019-56976-W;SUBJMETA.

[67] F. Shahamatifard, D. Rodrigue, F. Mighri, Thermal and mechanical properties of carbon-based rubber nanocomposites: a review, Plast. Rubber Compos. 52 (2023) 483–505. https://doi.org/10.1080/14658011.2023.2231286.

[68] M.A. Gimenes Benega, W.M. Silva, M.C. Schnitzler, R.J. Espanhol Andrade, H. Ribeiro, Improvements in thermal and mechanical properties of composites based on epoxy-carbon nanomaterials - A brief landscape, Polym. Test. 98 (2021) 107180. https://doi.org/10.1016/J.POLYMERTESTING.2021.107180.

[69] B. Le, J. Khaliq, D. Huo, X. Teng, I. Shyha, A review on nanocomposites. Part 1: Mechanical properties, J. Manuf. Sci. Eng. Trans. ASME 142 (2020). https://doi.org/10.1115/1.4047047.

[70] T. Stern, G. Marom, Fracture Mechanisms and Toughness in Polymer Nanocomposites: A Brief Review, J. Compos. Sci. 8 (2024) 395. https://doi.org/10.3390/jcs8100395.

Chapter 5

MICROSCOPIC AND OTHER CHARACTERISATION OF MATERIALS USED IN CARBON DIOXIDE CAPTURE

Abstract: Materials with effective capabilities for carbon dioxide capture and storage (CCS) have been the subject of intensive research due to the pressing need to reduce the rising amounts of CO_2 in the atmosphere. Understanding the morphology, structure, and surface characteristics of these materials at the micro- and nanoscales is a crucial part of this research. This chapter provides a thorough review of the microscopic and associated characterisation methods used to assess materials developed for CO_2 capture, with a focus on Transmission Electron Microscopy (TEM) and Scanning Electron Microscopy (SEM). Surface topography, pore distribution, and particle morphology are all thoroughly examined and explained by SEM, which also helps in evaluating surface area and textural characteristics that are crucial for adsorption processes. Conversely, TEM makes it possible to conduct high-resolution imaging and diffraction investigations, providing important insights into the dispersion of active sites within nanostructured materials, crystallinity, and lattice defects. SEM and TEM help to correlate structural features with CO_2 capture performance, as illustrated by case studies including metal-organic frameworks (MOFs), amine-functionalised mesoporous materials, and carbon-based adsorbents. In order to provide a comprehensive characterisation approach, complementary techniques like Fourier-transform infrared spectroscopy (FTIR), X-ray diffraction (XRD), and BET surface area analysis are also briefly discussed. The chapter demonstrates how the logical design and optimisation of next-generation CO_2 capture materials are directly supported by an understanding of microstructural and compositional factors through this integrated technique.

Keywords: Carbon Capture, SEM, TEM, CO_2

5.1 Introduction

There is a great deal of public anxiety about the possibility of a worsened climate as a result of global warming, as the CO_2 emissions are continuously rising and have reached up to 80%+ [1]. An increased reliance on burning fossil fuels (coal, oil, and natural gas) has been considered an important contributor to this sharp increase. These fuels are responsible for 86% of anthropogenic greenhouse gas emissions, with the remaining portion coming from chemical processing and changes in land use, mostly deforestation [2]. According to the IPCC, a contemporary conventional power plant with CCS technology could cut CO_2 emissions to the atmosphere by 80–90%. According to a recent estimate, for a typical coal-fired power plant, flue gas contains 12–15% CO_2 at 40 °C; the thermodynamic minimum energy penalty for removing 90% of the CO_2 from the gas is roughly 3.5% [3].

In contrast, traditional CO_2 capture with amine scrubbers will result in a 25-40% increase in a plant's energy needs. A higher electricity cost of $0.06 kWh or "avoided cost of capture" of $57–60/tonne CO_2 (as an alternative measure) has been projected by other recent cost analysis estimates based on near-term conventional regenerative amine scrubbing systems [4]. A growing number of fully integrated CCS projects are making it to the pilot and demonstration stages before going on sale. Compression, transportation (by existing pipelines), and storage of CO_2 are all technologically advanced and accessible components of the CCS process chain. Some smaller projects have started at different places across the globe such as the Dutch continental shelf (Netherlands), Snøhvit (Norway), La Barge (Wyoming, United States), Fenn Big Valley (Canada), Ketzin (Germany), and Schwarze Pumpe (Germany), in addition to the three large-scale demonstration projects that are currently in progress in Sleipner West (Norway), Weyburn (Canada), and Salah (Algeria). While one experiment at Schwarze Pumpe in Germany absorbs CO_2 at a coal-based plant, all of the current projects show carbon storage or reuse in enhanced coal-bed methane recovery methods [4,5].

Conventional methods of separating CO_2, such as solvent-based absorption, cryogenic distillation, and membrane-based separation, are still hampered by low regenerability, high-energy costs, amine losses, and low efficiency at low concentrations [6]. In order to effectively collect CO_2 effluents, a number of adsorbents and strategies have emerged over time. Various solid sorbent classes **(Figure 5.1)** that engage in physisorption and/or chemisorption and have demonstrated potential as effective CO_2 capture media; nevertheless, there are still many factors that need to be optimised [7]. Surface area, pore geometry and distribution, surface chemistry, functional group interactions, and structural configuration all affect how well these materials trap CO_2. These factors are continuously adjusted to maximise performance [7]. To further improve the capture capabilities, functionalisation or doping of the pore and sorbent surfaces may result in additional interactions with CO_2, such as electrostatic, carbamate bonding, H-bonding, and acid-base interactions. The adsorbent's chemical makeup, surface functional groups, polarisability, and CO_2 quadrupolar moment all influence the nature of these interaction forces. The design, synthesis, and development of such solid sorbents, as well as the optimisation of processes to maximise their output in terms of efficiency, economy, and energy, have thus been the focus of significant research resources during the past 20 years [8].

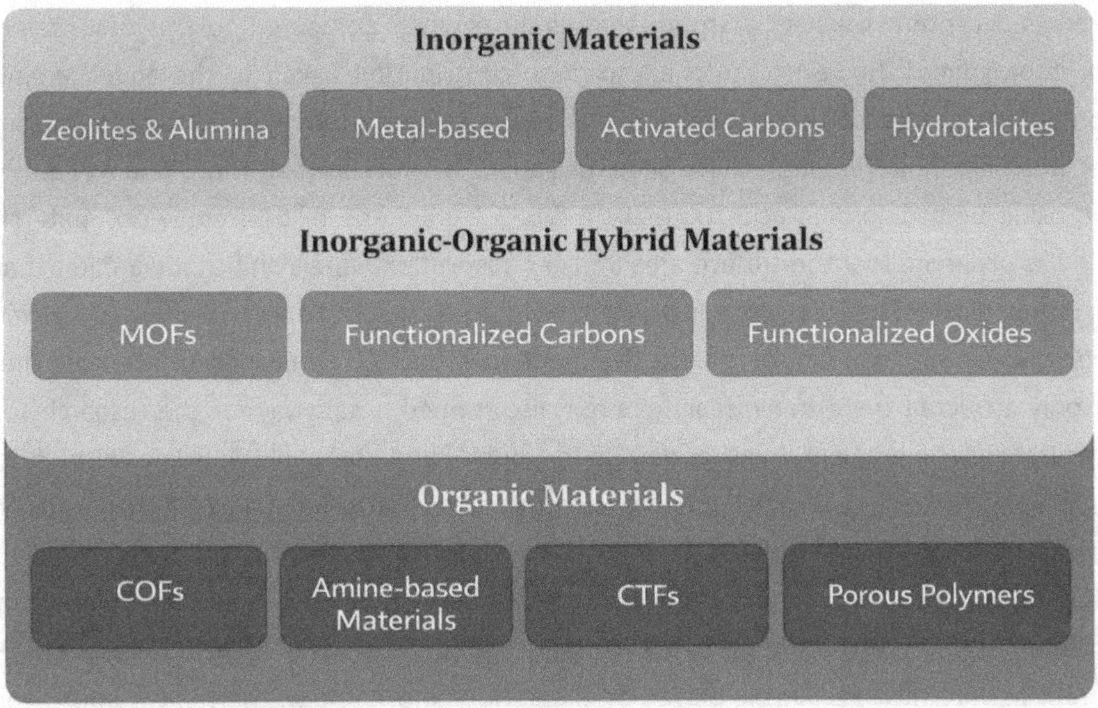

Figure 5.1: Illustration of different classes of materials used for Carbon Capture. Copyright Credit: (Ref. [7], Materials Today, Elsevier 2022)

Because of their great thermal and chemical stability, pure inorganic porous materials, such as zeolites, alumina, and mesoporous silica, have been thoroughly investigated for CO_2 sequestration and exhibit great promise [9, 10]. However, major barriers to their broad application are their high cost and structural rigidity [11]. CO_2 can bind to basic metal sites, particularly those with a strong ionic character, such as some alkaline and alkaline earth metal oxides, because of its Lewis acidity, forming mono- or multidentate species [12]. Lithium zirconates, calcium and magnesium oxides are the most widely utilised oxides in this context, albeit their kinetics may restrict their practical applications.

Nowadays large number of efforts are being put into finding very effective adsorbents. Two types of chemo sorbents that are currently being researched include silica-based materials and carbon-based materials that have been amine-functionalised. As CO_2 adsorbents, Yamin Liu and colleagues [13] produced KIT-6 loaded with tetraethylenepentamine. They discovered that when the concentration of TEPA at 60°C increases, it directly enhances the dynamic adsorption capacity as well. Over the course of 40 cycles of adsorption/desorption, the dynamic adsorption capacity decreased marginally (by roughly 5%). Serna-Guerrero *et al.* [14, 15] have also closely examined the CO_2 adsorption/desorption character and observed that triamine-grafted pore-expanded mesoporous silica is a suitable adsorbent for the removal of acid gas.

Additionally, it was shown that TRI-PE-MCM-41 remained stable for 100 adsorption–desorption cycles when the regeneration circumstances were appropriate. In order to separate CO_2, Plaza *et al.* [16] employed activated carbons loaded with amines. A sharp drop in microporous volume was attributed to the reduced capture capacity at room temperature when compared to the raw materials that were not impregnated. However, the chemisorption contribution linked to the added amino groups tends to enhance the carbon's performance at medium temperatures (70–90°C). The gas adsorption characteristics of graphite/graphene have been the subject of much investigation. Yamin *et al.* [13] coupled graphene oxide with diboronic acid to create a type of graphene oxide framework material. In the desired it was observed that the surface area and gas absorption were significantly greater than that of pristine graphene oxide. Density-functional calculations were used by Cabrera-Sanfelix [17] to examine the adsorption and reactivity of CO_2 on a defective graphene sheet. They proposed a tenable mechanism that not only attracted the other researchers but also framed a pathway for the research to address the same atmospheric problem on other defects in graphite and graphene. Bourlinos *et al.* [18] created a lightweight derivative that resembled graphite oxide and was foam-like. It was highly porous and had a large surface area. Low affinity for N_2, CH_4, and H_2 was demonstrated by the carbogenic foam's selective absorption of CO_2. The discussed aspects reveal the ongoing efforts in analysing the different materials, their efficiency, stability, and other properties that are the prerequisites for the CCS technology. Among the different important aspect's role played by the surface morphology, thermodynamic involved and other aspects will be discussed in this chapter.

5.2 Role of FESEM Studies in CO_2 Capture

A concentrated stream of high-energy electrons is employed in Field Emission Scanning Electron Microscopy (FESEM), a potent imaging method, to produce secondary electrons on a sample surface. This makes it possible to visualise surface topographies with lateral resolutions that are usually between 1 and 20 nm and a high depth of field. With magnifications up to approximately 100,000x, traditional scanning electron microscopes (SEM) that use thermal emission sources like tungsten filaments are suitable for general imaging tasks. However, a cold field emission (FE) cannon is recommended for applications that need higher resolution, such as the morphological assessment of sophisticated materials for CO_2 capture. By applying a powerful electric field at a sharp tip to extract electrons, FESEM systems with a cold FE source produce a substantially smaller beam size and higher electron brightness. Magnifications up to 500,000 times with improved spatial resolution are possible to attain through this technique [13, 16, 17]. One of FESEM's main benefits is its capacity to function at incredibly low accelerating voltages, which eliminates the need for conductive coatings and enables high-resolution imaging of non-conductive materials like porous adsorbents or polymeric membranes. FESEM is essential for examining the integrity, pore structure, and surface morphology of membranes, adsorbents, and catalysts in the context of CO_2 capture studies. It is an essential technique for the development and optimisation of materials intended for selective CO_2 separation because it aids in the visualisation of nanoscale characteristics that control gas diffusion and adsorption efficiency.

5.3 Nanostructured Carbon Nitrides' Role in CO_2 Capture

The prerequisites for any adsorbent to show promise for CO_2 capture are a highly porous and stable structure, high total pore volume, and a high specific surface area. An additional benefit for CO_2 collection under post-combustion CO_2 capture settings is the presence of ultra-micropores. It has been demonstrated that nitrogen-based structures like pyridine or pyrrolidine are advantageous for improving the interactions between the adsorbent and CO_2. A substance rich in nitrogen, carbon nitride, has been investigated for its potential in several fields, such as photocatalysis, electrochemistry, and catalysis. Since it exhibits the fundamental properties that establish its probability as a favourable material for CO_2 adsorption, the presence of nitrogen in the framework and, of course, on the external framework in the form of NH_2 groups in carbon nitrides is a significant advantage. When paired with a highly porous structure, this unusual characteristic provides the next generation of CO_2 capture adsorbents. Because of the presence of Lewis basic sites and a porous structure with high specific surface areas, carbon nitrides with varying nitrogen levels and defect sites always behave well toward CO_2 adsorption [19].

Nevertheless, bulk carbon nitride has very little surface area and is a comparatively non-porous structure. Because of their efficient CO_2 capture capabilities, particularly under high pressure, mesoporous carbon nitride nanostructures have recently emerged as a novel class of material that sheds light on the subject of CO_2 capture. However, the class of materials has not been explored and understood to a great extent regarding how the capturing mechanism works and its correlation with structure. Much greater attention is required to explain the presence of microporous carbon nitride materials for CO_2 capture, the synthesis of microporous carbon nitride, the properties they exhibit, and their correlation with CO_2 capturing, along with the responsible mechanism. For post-combustion CO_2 capture, which often occurs at ambient settings, a high microporosity in the adsorbent is a highly sought property. But under low-pressure circumstances, the special nitrogen present in carbon nitride acts as a helpful agent in the extraction of CO_2.

5.4 Prior Combustion Studies: CO_2 Capture

In contrast to other CO_2 adsorbents documented in the literature, the potential of nanostructured carbon nitrides for post-combustion CO_2 capture has received the least attention, as mentioned. However, there are examples of carbon nitride being developed into composite or hybrid nanostructures in the literature, and they have been explored in light of their current state of development. For instance, at 0.93 bar and 23 °C, melamine-based g-C_3N_4 nanosheets with SSA ≈ 35 m^2g^{-1} adsorbed a very small amount of CO_2 (0.18 mmol g^{-1}); this amount, however, increased significantly when either BIF-20 boron imidazolate (0.76 mmol g^{-1}) or ZIF-8 zeolitic imidazolate (1.26 mmol g^{-1}) structures were incorporated for the composite formation. When the corresponding composite nanostructures, BIF-20@ g-C_3N_4 (283 m^2g^{-1}), and ZIF-8@g-C_3N_4 (1178 m^2g^{-1}) were developed, the initial surface area of the porous BIF-20 (378 m^2g^{-1}) and porous ZIF-8 (1276 m^2g^{-1}) was somewhat decreased. Significant alterations in the original precursor's shape have also been observed along with the variations in the porosity [20]. These nanostructures have a strong potential for post-combustion CO_2 capture because of their significantly greater capacity to trap CO_2. It has also been claimed that carbon nitride's ability to adsorb CO_2 can be improved by physically impregnating it with solvents based on amines. To improve the CO_2 adsorption capabilities, for instance, g-C_3N_4 made from urea was functionalised by adding varying concentrations of polyethyleneimine (PEI) to its stacked pores by employing the solvent-evaporation technique [21].

In comparison to pristine g-C_3N_4 (0.25 mmol g^{-1}), it has been observed that g-C_3N_4 supports PEI on its surface. Additionally, the presence of amine groups on the surface of the composite PEI@g-C_3N_4 nanostructures increased the adsorption of CO_2 at 25°C by 6.8 times (1.71 mmolg^{-1}). It is important to emphasise that the presence of amine groups integrated on the surface is the only reason for any CO_2 absorbed, and the specific surface area of the composite nanostructures has no bearing on CO_2 adsorption.

Additionally, the accessibility of the surface amine groups by CO_2 is the reason why the CO_2 uptake reduces as the temperature rises up to 50 °C and then begins to increase at temperatures above 50 °C–100 °C [21]. The above discussion clearly highlights how crucial it is to functionalise the nanostructured carbon nitrides in order to maximise the low-pressure adsorption of CO_2 on their modified surface with improved physico-chemical characteristics. Furthermore, such composites' high specific surface area would help to improve CO_2 adsorption on a much greater scale. To enable high CO_2 capture under post-combustion conditions, the next generation of carbon nitride materials must be created with highly developed porous architectures and high specific surface area in microporous domains.

A good catalytic activity for Knoevenagel condensation was demonstrated, for instance, by bulk carbon nitride made from ethylenediamine, carbon tetrachloride, and templated into microporous structures using MCM-22 [22]. The carbon nitrides' microporous structure and basicity were both credited with the catalytic activity, which was also demonstrated to be much superior to that of their mesoporous counterparts. Carbon nitride nanostructures have shown success in a variety of applications, such as sensing and imaging, attributed to their fundamental characteristics and perfect electrical structure [23]. The intriguing electronic characteristics of carbon nitride may aid in the rationalisation of the upcoming generation of adsorbents, which will allow for the capture and release of CO_2 in a single system by simply adding or removing charge on conductive carbon nitride structures like g-C_3N_4 [24]. In addition to making simultaneous CO_2 collection and release easier, the voltage change enables selective CO_2 adsorption in a mixture of CH_4, H_2, and N_2. More significantly, this electrocatalytic switching technology can make the adsorbent moisture-free. Designing extremely efficient carbon nitride nanostructures for CO_2 capture is still a long way off. However, regarding the pre-combustion CO_2 capture employing carbon nitrate, there are a lot of reports related to the templating of the nitrate mesopore structures for CO_2 capturing under high pressure. A prominent report about this is the development of spherical mesoporous carbon nitride (MCN-7) by employing EDA with CCl_4 and FDU-12 as a template. The developed material, MCN-7, has been found to exhibit an efficient specific surface area of the order of 901 m^2g^{-1}. Attributed to this, along with porous and ordered nanostructures of the material, they have demonstrated an adsorption capacity of 13.5 mmol g^{-1} [25]. Knowing the fact that the specific surface area as well as porosity of the materials under discussion is very low, the discussion depicts that the templating approach could be a viable option for future efficient materials as an adsorbent for CCS technology, while dealing with the pre-combustion process.

5.4.1 Role of BET Studies: CO_2 Capture

In order to assess the porosity properties of carbon materials, nitrogen (N_2) adsorption–desorption isotherms at 77 K are commonly measured using a volumetric adsorption analyser [26]. Adsorption isotherms can be divided into six categories as per the IUPAC classification, with Type I or Type IV isotherms being most frequently found in porous carbon materials [27]. **Figure 5.2** shows examples of these types of isotherms. The degree of interaction between the sorbate and the adsorbent surface, along with the existence, size, and distribution of pores, determines the classification. All samples do not rigidly follow a single isotherm type. The co-existence of several pore structures occasionally results in the emergence of hybrid isotherm profiles. For example, Type I and Type II, or Type I and Type IV isotherms, are frequently present in porous carbons with high specific surface areas, suggesting the presence of both micropores and mesopores [28]. A Type I isotherm, which is typical of microporous surfaces, is generally indicated by a fast rise in adsorption at low relative pressures ($p/p_o < 0.1$) without any hysteresis loop over a broad pressure range [29]. Accordingly, the microporous character of these samples is confirmed by the pore size distribution (PSD) curves, which primarily display pores smaller than 2 nm. On the other hand, mixed Type I and Type IV adsorption–desorption isotherms are commonly seen in samples that were generated using distinct chemical activating agents or at different temperatures. When mesoporous or macroporous structures are present in the carbon matrix, an H1-type hysteresis loop will occur at higher relative pressures ($p/p_o > 0.5$). The formation of distinct cylindrical holes or interconnected pore networks is suggested by this behaviour. Furthermore, it has been observed that a shift in isotherm behaviour from Type I to Type IV occurs when the activation time during synthesis is increased, which raises the mesopore volume (Vmeso) in the carbon structure [30]. Non-porous carbons, such as biochar or ash-rich carbon compounds, have a very limited potential to absorb nitrogen (N_2) because of the preponderance of ultra-micropores. The specific surface area (S_{BET}) derived from N_2 adsorption isotherms is dramatically reduced at cryogenic temperatures due to the considerable inhibition of the diffusion rate of N_2 molecules into ultra-micropores. S_{BET} values in these situations usually range from 0 to 10 $m^2 \cdot g^{-1}$ [31]. An appreciable increase in N_2 uptake indicates that the porosity of carbon materials improves significantly as the pyrolysis temperature rises. The activating chemicals' etching impact during heat treatment, which promotes the formation of microporous frameworks, is primarily responsible for this improvement. Both S_{BET} and total pore volume increase quickly with increasing pyrolysis temperature, indicating a more accessible porous network [32]. The breakdown of mesopore walls, which leads to the formation of larger-sized micropores, is thought to be the etching mechanism. However, pore collapse can occur when the carbonisation temperature is raised over a specific point, especially around 1200°C, which significantly reduces the porosity overall. Excessive activation and the pore wall's lack of mechanical integrity at extreme heat are considered to be responsible for this reduction. Additionally, elemental doping or surface changes can lower the observed S_{BET} because they introduce structural flaws that disrupt the pore network or partially block pores. Char is left behind when biomass breaks down during pyrolysis, releasing volatile components. The residual carbon goes through additional gasification as the temperature rises above 700°C. During this phase, the Boudouard reaction is crucial:

$$CO_2 + C \leftrightarrow 2CO \ \Delta H = +175.52 \ kJ \cdot mol^{-1}$$

Because of this endothermic reaction, which etches the solid matrix when CO_2 combines with carbon to generate CO gas, porosity continues to develop. Both the activation time and temperature affect the rate and magnitude of the reaction. Pore formation rises as activation time increases because more carbon is used. But at extremely high temperatures (1000–1200 °C), the Boudouard reaction's equilibrium constant drastically decreases, moving the reaction closer to equilibrium and preventing further pore formation. Under such circumstances, the observed drop in S_{BET} is attributed to this shift through gasification [33, 34].

5.5 Isotherms Modelling

The term "adsorption phenomenon" refers to a thermodynamic equilibrium process that takes place at the interface between two phases, usually a solid (adsorbent) and a gas (adsorbate), under constant pH and temperature [35]. The equilibrium state denotes a balance between the concentration of residual adsorbate in the gas phase and the amount of gas adsorbed. Important information regarding the adsorption mechanism and the surface properties of the adsorbent can be gleaned by examining the physicochemical parameters controlling this equilibrium. For gas separation applications, such as CO_2

collection, this data is essential for the design, modelling, and deployment of adsorption systems [36]. There are four basic steps in the mechanism of CO_2 adsorption on solid materials, according to both theoretical and experimental research: (i) the molecules of CO_2 make contact with the adsorbent surface; (ii) the molecules penetrate and diffuse into the internal pore structure; (iii) the molecules interact with active adsorption sites; and (iv) the gas accumulates or binds within the available micropores or surface regions [37]. On the other hand, there are several adsorption isotherm models that have been created to interpret and quantify these processes. The equilibrium adsorption capacity (q′) on the adsorbent surface and the equilibrium concentration (C′) of the adsorbate in the gas phase are related in these models and are represented through different mathematical equations. The adsorption of CO_2 on commercially available activated carbon made from lignite granules was investigated by Singh and Kumar. In contrast to the Langmuir model, their findings showed a superior fit with the Dubinin-Astakhov (DA) isotherm model [38]. In a similar vein, Kim *et al.* used K_2CO_3 activation to create activated carbon from wasted coffee grounds and assessed its CO_2 adsorption capabilities at 0, 25, and 50°C. At every measured temperature, the experimental data have been found to closely match the Langmuir isotherm [39].

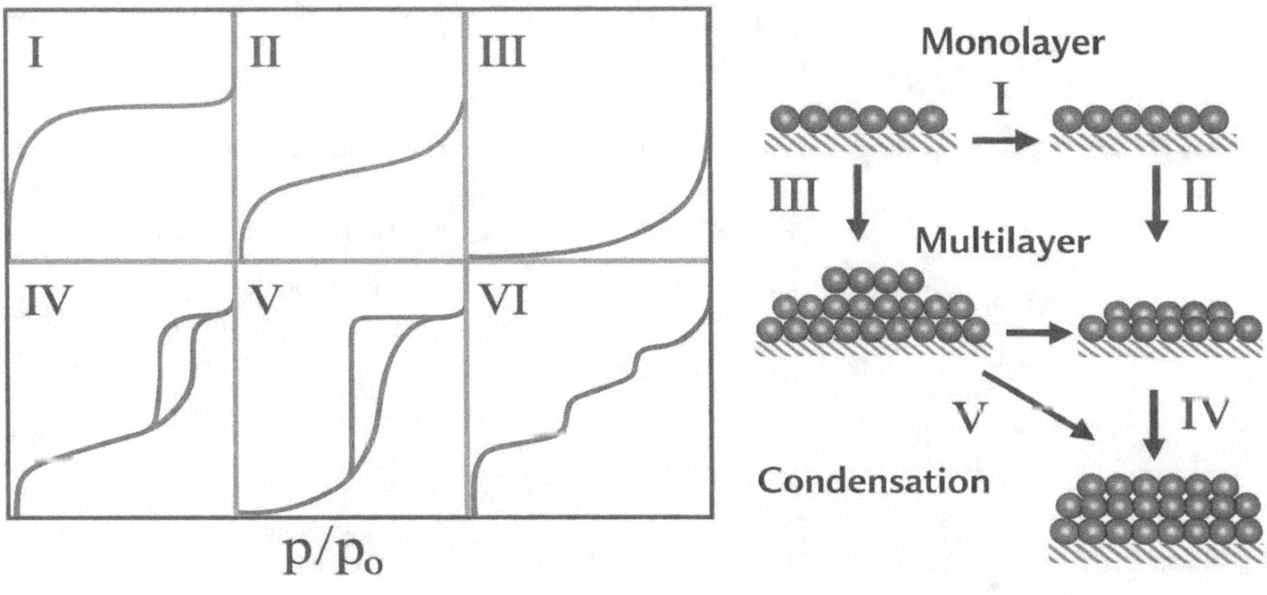

Figure 5.2: Types of adsorption isotherms based on the International Union of Pure and Applied Chemistry (IUPAC). Copyright Credit: (Ref [41], Advanced Science, Wiley, 2023)

Table 5.1: Different models of Isotherm Modelling

Various Isotherm Models	Non-linear Forms	Linear Forms	Model Parameters	Ref.
BET	$q_e = \frac{q_s C_{BET} C_e}{(C_s - C_e)\left[1+(C_{BET}-1)\left(\frac{C_e}{C_s}\right)\right]}$	$\frac{C_e}{q_e(C_s - C_e)} = \frac{1}{q_s C_{BET}} + \frac{(C_{BET}-1)}{q_s C_{BET}}\frac{C_e}{C_s}$	q_s: Max adsorption capacity (mg/g); C_s: Monolayer saturation conc. (mg/L); C_{BET}: BET constant (L/mg)	[42]
Freundlich	$q_e = K_F C_e^{\frac{1}{n}}$	$\log q_e = \log K_F + \frac{1}{n} logC_e$	K_F: Freundlich constant; n: Adsorption intensity	[43]
Langmuir	$q_e = \frac{Q_0 b C_e}{1+bC_e}$	$\frac{C_e}{q_e} = \frac{1}{bQ_0} + \frac{C_e}{Q_0}$	Q_0: Max adsorption capacity (mg/g); b: Langmuir constant (L/mg)	[44]
Temkin	$q_e = \frac{RT}{b_T}\ln A_T C_e$	$q_e = \frac{RT}{b_T}\ln A_T + \frac{RT}{b_T}\ln C_e$	b_T: Temkin constant (J/mol); A_T: Binding constant (L/g	[45]
DA	$q_e = q_{m,DA}\left[-\left(\frac{\varepsilon_{DA}}{E_{DA}}\right)^{n_{DA}}\right]$ $\varepsilon_{DA} = RT\ln\left(\frac{C_{s,DA}}{C_e}\right)$	$lnq_e = lnq_{m,DA} - n_{DA} ln\frac{\varepsilon_{DA}}{E_{DA}}$	$q_{m,DA}$: Max adsorption (mg/g); n_{DA}: Heterogeneity; E_{DA}: Energy (kJ/mol)	[46]
DR	$q_e = (q_m DR)exp^{(-k_{ad}\varepsilon^2)}$ $\varepsilon = RT\left(1+\frac{1}{C_e}\right)$	$\ln q_e = \ln q_s - K_{ad}\varepsilon^2$	$q_{m,DR}$: Max adsorption (mg/g); k_{ad}: DR constant (mol^2/kJ^2)	[47]
Halsey	$q_e = \left(\frac{K_H}{C_e}\right)^{\frac{1}{n_H}}$	$\ln q_e = \left(\frac{1}{n_H}\right)\ln K_H - \left(\frac{1}{n_H}\right)\ln C_e$	K_H, n_H: Halsey constants	[48]
Henry		$q_e = K_{HN} C_e$	K_{HN}: Henry adsorption constant ($L\,g^{-1}$)	[49]
Jovanovich	$q_e = q_{max}\left(1-e^{K_J C_e}\right)$	$\ln q_e = \ln q_{max} - K_J C_e$	q_{max}: Max adsorption (mg/g); K_J: Jovanovich constant (L/g)	[50]
MET	$q_e = q_s\left[\frac{k}{ln\left(\frac{C_s}{C_e}\right)}\right]^{\frac{1}{3}}$		q_s: Saturation capacity (mg/g); k: MET constant; C_s: Saturation conc.	[51]
Radke-Prausnitz	$q_e = \frac{a_{RP} r_R C_e^{\beta_R}}{a_{RP} + r_R C_e^{\beta_R - 1}}$		a_{RP}: Max adsorption (mg/g); r_R: RP constant; β_R: Exponent	[42]
Sips	$q_e = \frac{q_{m,s} K_s C_e^{\beta_s}}{1+K_s C_e^{\beta_s}}$	$\beta_S \ln(C_e) = -\ln\left(\frac{K_S}{q_e}\right) + \ln(a_s)$	$q_{m,s}$: Max adsorption (mg/g); Ks: Sips constant (L/g); βs: Exponent	[52, 53]

5.6 CO_2 Capture: Advancements and Goals

The phenomenon of CO_2 Capturing has gained attention across the globe and is being carried out at different stages of progress. It involves different stages, among which a few have progressed to a great extent, a few are at intermediate stages of progress, and a few need proper attention to initiate their progress. The different stages involved in this process are well depicted in **Figure 5.3.**

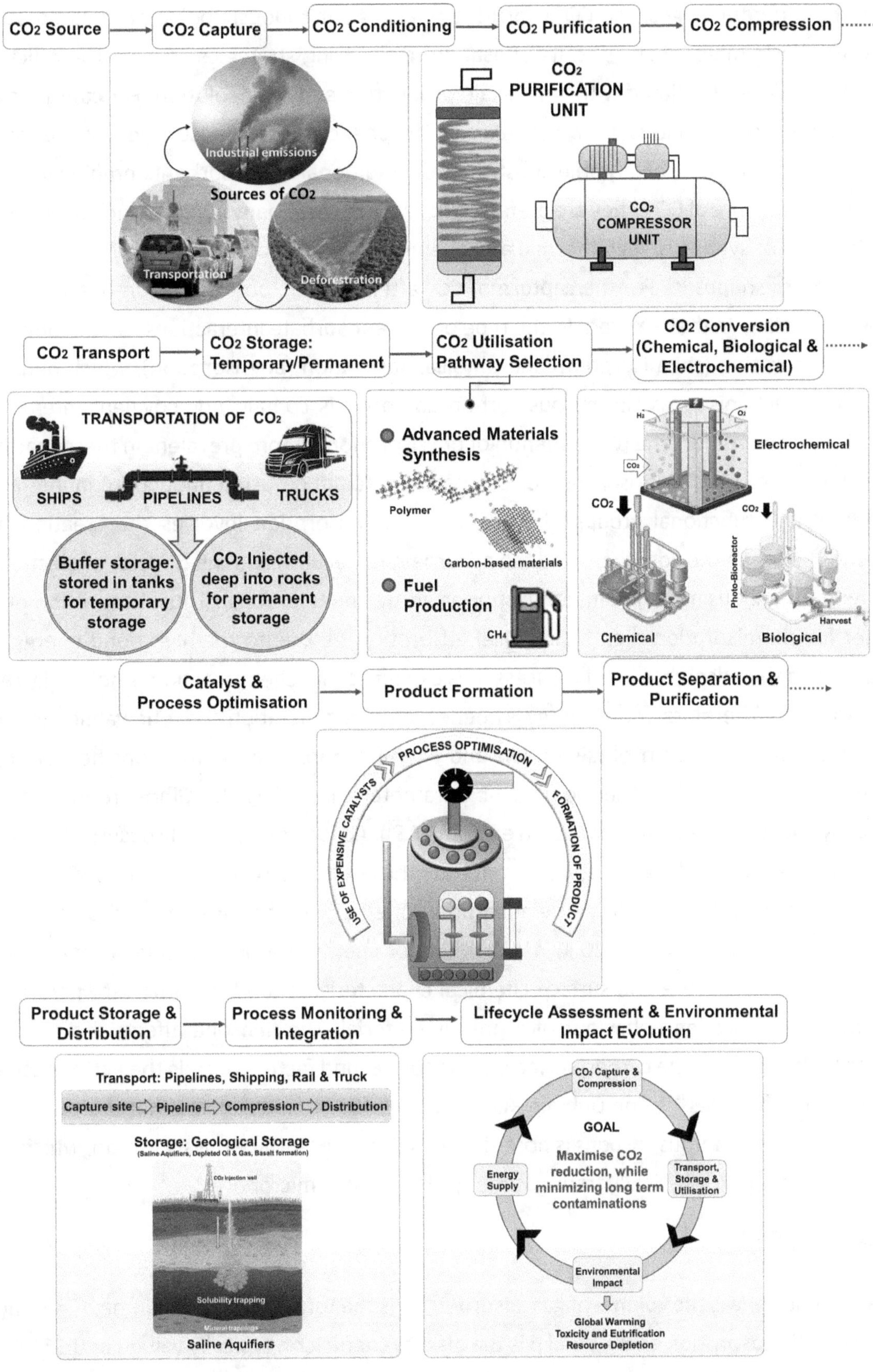

Figure 5.3: General overview of the complete CCUS framework, starting from the very basic anthropogenic CO_2 source, to capturing and processing units, followed by the transport, storage, and conversion pathways. Copyright Credit: (Ref. [3], Energy Environmental Science, RSC, 2026)

Among the different stages, the CO_2 capturing phase is continuously being progressed at different levels from the basic material stage to its utilisation, developing strategies to tune the efficiency, and finally, the implications associated with large-scale production and their solution. For capturing different materials, analysis is being conducted, which relies on the phenomenon of adsorption. In this regard, solid adsorbents are considered to be a viable substitute for traditional liquid sorbents employed in chemical absorption for the capture of CO_2. They are preferred because of their many benefits, including affordability, sustainability, ease of synthesis, high thermal and chemical stability, accessibility, surface hydrophobicity, and adaptable pore shapes [54]. For the capture of CO_2 after burning, porous carbon-based materials are particularly interesting. In these materials, the type of sorbent-sorbate interactions can be roughly divided into chemical adsorption (chemisorption) and physical adsorption (physisorption), which determine the CO_2 adsorption mechanism. As the porous carbon compounds contain non-covalent interactions and energy-efficient regeneration, the phenomenon of physisorption is more prevalent in these materials. The general mechanism of physical adsorption of CO_2 occurs through diffusion (induction) in interconnected porosity or through functional groups [41]. Conversely, chemisorption involves the creation of strong covalent bonds between the adsorbent and the adsorbate molecules, and the transfer of electrons occurs between them (mainly using C-rich material supported by amine groups). Thus, adsorption enthalpy is much higher for chemisorption than for physical adsorption [55]. Due to the high bond energy between the adsorbent and the adsorbate in this process, it is expected that chemisorption is not easily reversible, unlike physisorption. Therefore, more energy is needed in the regeneration and renewability of the porous carbon-based phenomena. Both physisorption and chemisorption mechanisms on the heterogeneous surface can be distinguished by thermodynamic parameters, including the Gibbs free energy changes ($\Delta G°$), enthalpy changes ($\Delta H°$), and changes in entropy ($\Delta S°$). Adsorption generally occurs physically on the surface of porous carbon, making it easier to regenerate the adsorbent with minimal energy requirements. Parameters such as standard enthalpy (the adsorption of energy in the form of heat) and isosteric heat (Q_{st}) for physisorption are usually < 20 kJ mol^{-1}, while for chemisorption, it resides in the range of 40–200 kJ mol^{-1} [56]. Pore structure and surface chemical characteristics are two important parameters that influence the CO_2 adsorption behaviour of a porous material. The first one affects the contact forces between gas molecules and the carbon surface, while the second factor controls the active sites available for admitting CO_2 molecules [57]. The Dubinin–Radushkevich (D–R) model states that the primary process for CO_2 sorption in microporous carbons is pore-filling. In this case, surface physisorption, which is defined by the following equation, physically traps gas molecules in small micropores:

$$V=V_o \exp[-(A\beta E_0)^2]$$

In the above equation, V is the volume of gas adsorbed, V_o is the total micropore volume, $A = RT\ln(p/p_o)$, E_o is the characteristic adsorption energy, and β is the affinity coefficient, typically valued at 0.35 [58]. Through dispersion and induction forces, the quadrupole form of CO_2 molecules further improves their interaction with polar and ionic sites on the carbon surface. Increased surface concentration results from molecules attaching more readily to the adsorbent surface as their free energy decreases. Once equilibrium is achieved and the surface is saturated, multilayer adsorption or surface diffusion takes over [59].

5.7 CO_2 Adsorption Thermodynamics

One important thermodynamic metric for assessing adsorbent performance is the isosteric heat of adsorption (Q_{st}). It aids in determining the need for regeneration energy as well as adsorption strength. Furthermore, the variability of adsorption sites is reflected in Q_{st} [60]. The Clausius-Clapeyron equation is employed to calculate it from equilibrium isotherms:

$$-\frac{Q_{st}}{R}=\left(\frac{\partial lnP}{\partial(1/T)}\right)_n$$

where T is temperature (K), P is pressure (kPa), R is the universal gas constant (8.314 J mol^{-1} K^{-1}), and Q_{st} (kJ mol^{-1}) is the isosteric heat. Assuming that Q_{st} is temperature independent, integration produces:

$$\text{Ln P} = Q_{st}/RT + \text{constant}$$

An estimate of Q_{st} is obtained from the slope of the plot of ln$P$ vs $1/T$ [60].

5.7.1 Technologies for CO_2 Adsorption Processes

The adsorption of CO_2 from a gas mixture onto the surface of a solid adsorbent and the renewal of the adsorbent by desorbing the trapped gas are the two primary stages in most CO_2 adsorption methods. Numerous cyclic adsorption methods (a few depicted in **Figure 5.4**) have been employed for industrial use in order to accomplish these processes effectively.

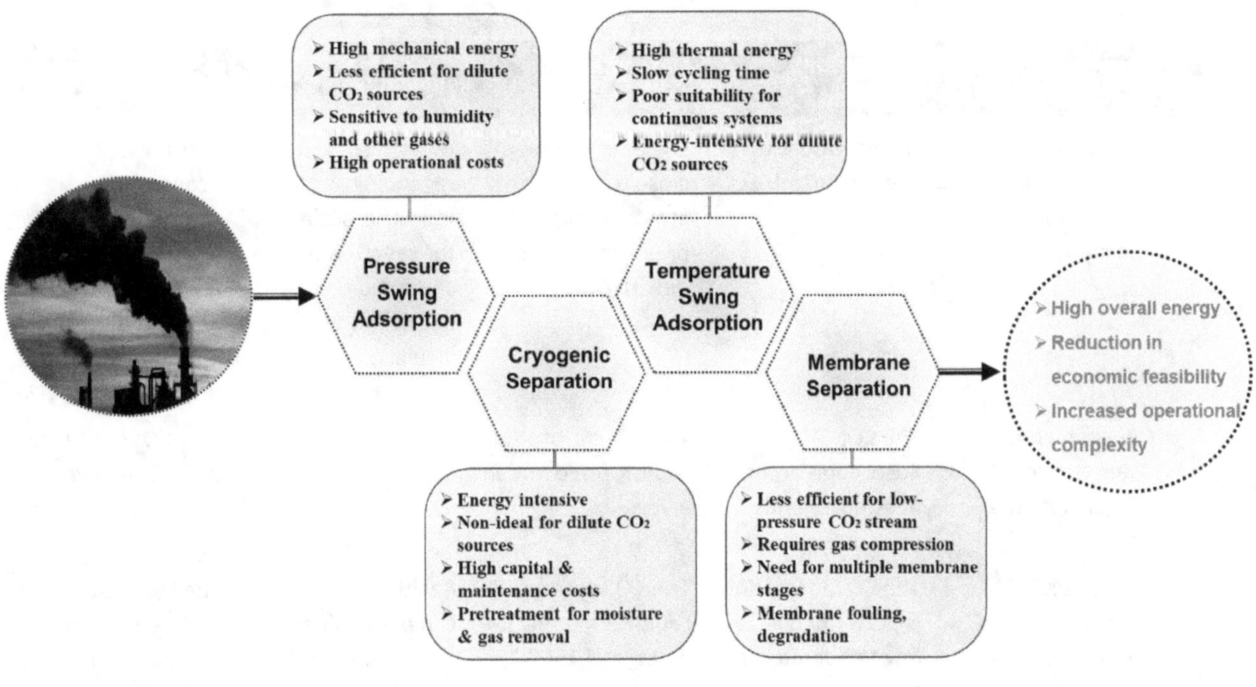

Figure 5.4: Schematic presentation of the different processes involved in the CCUS technology for CO_2 capturing and separation, along with the operational considerations and economic implications. Copyright Credit: (Ref. [3], Energy Environmental Science, RSC, 2026)

Among these, Pressure Swing Adsorption (PSA) provides a useful and scalable technique for bulk separation by adsorbing CO_2 at high pressure and releasing it at low pressure [61]. Conversely, VSA uses a vacuum for the desorption step and operates at near-atmospheric pressures, which makes it energy-efficient and appropriate for post-combustion applications [62]. Although it tends to be more energy-intensive because of its heating requirements, TSA uses steam or thermal input to replenish the adsorbent by altering the adsorption–desorption equilibrium. A more recent innovation is Electrical Swing Adsorption (ESA), which uses electrical energy to cause localised heating that initiates desorption [62, 63]. The great efficiency, quick cycle times, and reduced operating costs of PSA and VSA make them the most popular among them. For large-scale CO_2 capture, PSA is particularly preferred, but VSA offers a more energy-efficient option for systems running in ambient settings [63].

5.7.2 Material Requirements and Selection Challenges

Choosing the right adsorbent is essential for effective CO_2 capture. High selectivity for CO_2 over competing gases like N_2, high adsorption capacity under practical operating circumstances, fast adsorption–desorption kinetics, strong regenerability over several cycles, and overall economic viability are all characteristics of an ideal material. Rajagopalan *et al.* [64] have pointed out that great selectivity is insufficient on its own. Due to the requirement for bigger adsorbent beds or frequent regeneration cycles, a material with high selectivity but poor adsorption capacity may make the process ineffective and expensive.

Table 5.2. Summary of various parameters of PSA, VSA, and TSA [66].

Technique	Principle	Adsorption conditions	Desorption conditions	Advantages	Disadvantages
PSA	Relies on pressure variation to regenerate the sorbent.	High pressure is used to adsorb CO_2	Lower pressure, close to atmospheric pressure, is used to desorb CO_2.	Most used technology at the industrial scale for gas stream separation, due to its ease of use and viability.	May not be economical to compress CO_2 along with N_2 in real flue gas conditions. Sorbent selectivity is low at high pressures.
VSA	Relies on pressure variation to regenerate the sorbent.	The CO_2 adsorption occurs at atmospheric pressure	The CO_2 desorption is performed under a vacuum		Low CO_2 purity and low CO_2 recovery
TSA	Relies on temperature variation to regenerate the sorbent.	The CO_2 adsorption is performed at low temperatures to maximise adsorption.	The CO_2 desorption requires heating in a range of 150–200°C.	Relatively pure CO_2 can be obtained, as compared to other techniques.	The high heat required for sorbent regeneration makes the process expensive.

Consequently, process-level factors need to be taken into account while choosing adsorbents. To verify the materials' performance in the actual world, operating conditions, including cycle duration, pressure, and temperature, must be simulated and optimised as well. Developing adsorbents with greater capacity and selectivity, improving adsorption processes with these materials, and combining adsorption modelling with techno-economic analysis to evaluate the system's overall viability are the three main components of the most successful method [65]. **Table 5.2** provides a summary of comparative performance statistics for PSA, VSA, and TSA technologies [66].

5.8 Specifications Affecting CO_2 Adsorption

Numerous intrinsic and extrinsic factors affect the adsorption capacity of porous carbon materials generated from biomass [67]. The volume and distribution of micropores and ultra-micropores, pore ordering, the existence and kind of surface functions, and the presence of heteroatoms like nitrogen or oxygen are examples of intrinsic variables. Adsorption behaviour is also significantly influenced by properties such as the oxygen-to-nitrogen (O/N) ratio, dipole interactions, hydrogen bonding capabilities, and defect configurations. However, the adsorption performance is also directly impacted by external variables, such as temperature, adsorption pressure, moisture content, and the presence of co-adsorbed gases. Although total pore volume and specific surface area (SSA) are often used metrics to evaluate adsorption potential, several researchers contend that these metrics are inadequate on their own. Rather, when it comes to improving CO_2 retention capacity, the kind and accessibility of ultra-micropores, especially those smaller than 0.7 nm, are frequently more important [68]. The ultimate pyrolysis temperature, gas flow rate during activation, acid pickling, and residence duration are examples of synthesis variables that have a substantial impact on the surface chemistry and resulting textural characteristics of the porous carbon, which in turn affect its CO_2 collection efficiency. Therefore, adjusting these parameters is crucial to maximising the effectiveness of carbon-based adsorbents in actual CO_2 capture situations.

5.9 Parameters Affecting CO_2 Adsorption

The adsorptive capability of porous carbon depends on several parameters. Some parameters directly affect the adsorption quantity, and others indirectly. The meaning of indirect parameters is that some synthesis conditions, such as a change in pyrolysis temperature, the flow rate of gases during synthesis, acid pickling, and retention time of the sample at the final pyrolysis temperature, affect the high SSA, which eventually determines the total adsorption capacity. However, it has been observed that the high SSA and the total volume of porosity obtained from changing the synthesis conditions do not have much effect on gas adsorption, but the type of porosity resulting from the synthesis method does play an effective role in trapping gas molecules [68]. Two determinants influence CO_2 adsorption: intrinsic factors and external parameters. The order of porosity, the volume of micropores and ultra-micropores, the presence of heteroatoms (intrinsic and doped), hydrogen bonds, the type of nitrogen, the ratio of O/N, surface, and functional groups are all intrinsic variables. On the other hand, external factors such as moisture, temperature, and pressure of adsorption affect the final adsorption capacity of the adsorbents [68]. All these parameters are well presented in **Figure 5.5** and are briefly described as:

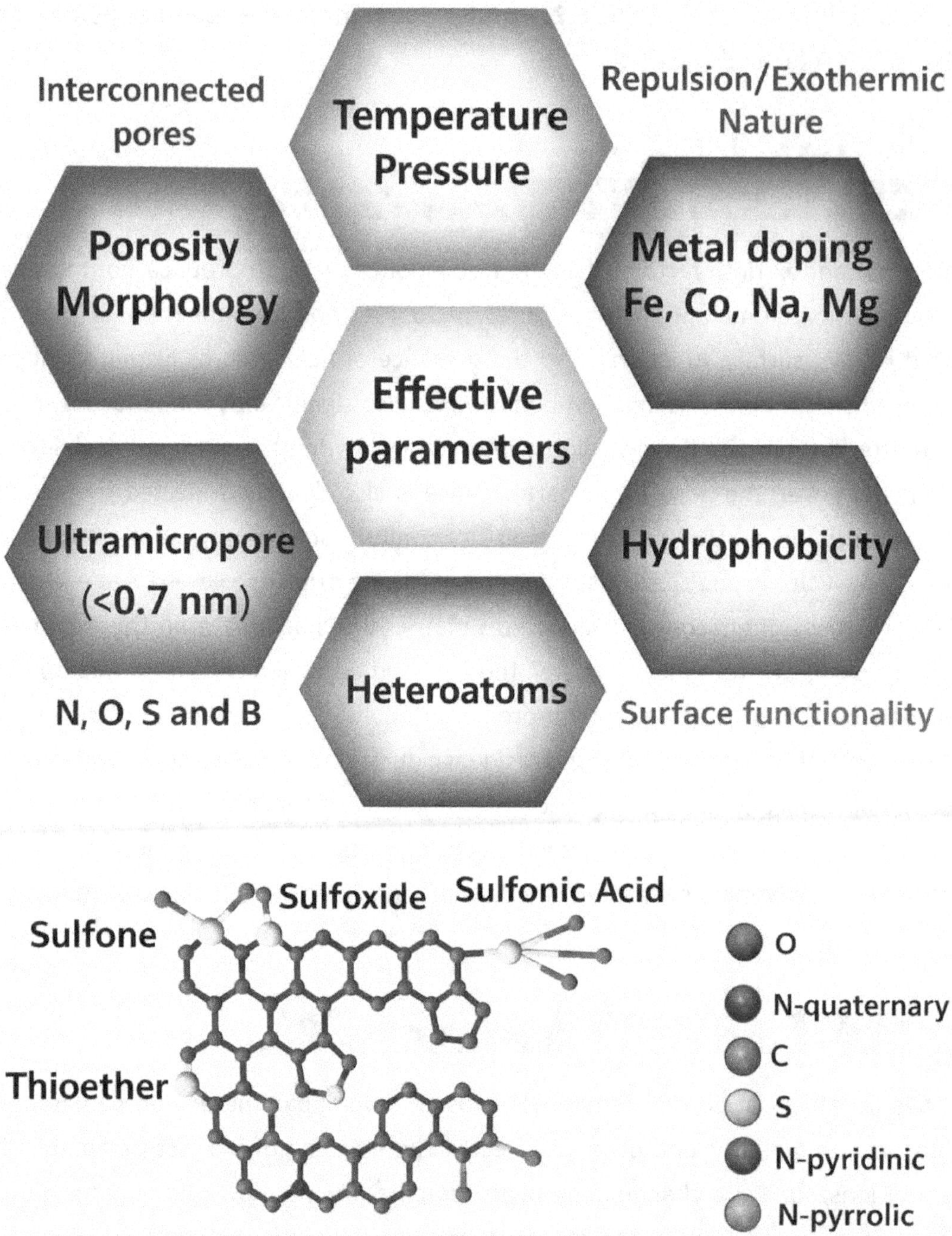

Figure 5.5: General factors affecting the absorption of CO_2 on various natural-derived porous carbon wastes. Copyright Credit: (Ref [41], Advanced Science, Wiley, 2023)

5.10 Porosity: In CO_2 Capture

Carbon frameworks with hierarchical porosity, which include macropores (> 50 nm), mesopores (2–50 nm), and micropores (< 2 nm), are especially well adapted for CO_2 uptake. By decreasing diffusion paths, this combination of pore sizes provides high specific surface areas and linked channels that promote effective molecular diffusion. The prevention of gas molecules from getting trapped in the blind pores is an effective approach for maximising the available surface area for adsorption, which is one of the main benefits of such linked porosity. The number of accessible active sites rises with the amount of space available for gas-solid interactions, which eventually improves CO_2 collection. Additionally, the channels of communication between various pore domains have a significant impact on adsorption kinetics [69]. Ordered mesoporous carbons (OMCs) have special pore properties that are advantageous for CO_2 collection, especially those with 2D hexagonal symmetry that are created using hard or soft templating techniques. These materials improve molecular diffusion and adsorption because they usually have a uniform surface and a well-organised configuration of mesoporous channels. Adsorption capacity is increased by the additional van der Waals contacts that are introduced between gas molecules and the carbon framework by the presence of micropores lining the mesopore walls. The well-ordered meso-structures in micro-mesoporous OMCs have better CO_2 transport and storage capabilities than disordered mesoporous or only microporous carbons. Recently, a non-toxic nitrogen-rich biomass precursor for the synthesis of N-doped ordered mesoporous carbons (NOMCs) like CMK-8, D-glucosamine, has been investigated [70]. With a capacity of 1.47 mmol g^{-1} at 30°C under 0.9 bar, these materials have been found to exhibit enhanced CO_2 adsorption, which is noticeably greater than that of undoped OMCs (0.97 mmol g^{-1} at the same conditions). Similarly, utilising hierarchical micro-mesoporous carbons made from coal tar under various synthesis circumstances, CO_2 absorption of 2.38 mmol g^{-1} and 2.18 mmol g^{-1} at 25 °C was attained [71]. Despite their benefits, the adsorption effectiveness of these porous carbons may be limited by their frequent lack of adequate microporosity. Carefully optimising synthesis procedures can address this to introduce micropores while preserving structural order. Large-scale industrial use may face difficulties due to the cost-prohibitive nature of templating techniques, which are essential for producing organised porosity. The advancement of these materials for practical CO_2 capture thus still faces a significant challenge: obtaining the required structural order and porosity at a reasonable cost [72].

5.11 Micropore Size: An Important Factor

The ultimate pore architectures can be significantly determined by the activation factors, including the activation temperature and the reagents. An enormous number of small mesopores, massive macropores, and narrow micropores can be produced by etching the structure of porous carbon with different chemicals. The results indicate that the linked macropore and microporous architectures ensure high sorption performance of porous materials. The macroporous structure (50 ≤ d ≤ 100 nm diameter) facilitates rapid gas movement, while the micropores (≤ 2 nm diameter) act as the critical adsorption sites [73]. An exceptional CO_2 sorption capacity was achieved for the ultra-micropore (≤ 0.7 nm diameter) structure developed, causing the interaction energy to increase considerably and the potential fields of adjacent walls to overlap. A smaller micropore volume frequently denotes the presence of more robust sorption sites in porous sorbents. Especially at low pressures, a material's ability to adsorb CO_2 is greatly influenced by its microporous properties. Due to their low kinetic energy, gas molecules have a significant propensity to go into the small gaps of micropores. CO_2 molecules usually adsorb onto the internal and external surfaces of these domains, resulting in the formation of a monolayer. Similarly, at high pressures (e.g., ~30 bar), small mesopores (usually 2–4 nm) play a major role in CO_2 uptake by offering more surface area and room for gas storage. Nevertheless, both micropores and tiny mesopores can create diffusional restrictions, which could hinder sorption and desorption kinetics, even though they are successful in gas adsorption. Furthermore, the adsorption rate may be further hampered by the strong capillary forces inside micropores [74].

5.12 Role of Heteroatoms: In CO_2 Capture

Following the optimisation of ultra-microporosity, surface modification and functional group engineering have become very successful methods for increasing the CO_2 adsorption capacity of porous materials. Amine functionalisation, physical and chemical activation, and heteroatom doping, especially with elements like nitrogen, oxygen, sulphur, and boron, have been the main areas of research (as discussed in the previous chapter). Nitrogen doping has drawn a lot of interest among them since it can affect gas uptake in several ways. There are three ways to look at nitrogen's role in CO_2 adsorption: as an inert gas during carbonisation, as an intrinsic or non-intrinsic source of nitrogen, and as the nitrogen functionalities that are generated. Because nitrogen contains five valence electrons and a similar atomic radius to carbon, its addition to the carbon matrix can change the material's conductivity, surface basicity, and electronic structure, depending on the doping technique and amount. At 0.15 pressure and 25°C, for instance, Dawei *et al.*, demonstrated that a modest nitrogen flow rate of 800 mL min^{-1} during activation improved CO_2 adsorption up to 1.86 mmolg^{-1} [75]. The material's surface polarity and basicity are increased by nitrogen's higher electronegativity than carbon, which improves the material's affinity for CO_2. Nitrogen doping also creates localised adsorption sites by altering the spin density and atomic charge distribution. Because of its linear geometry and C=O bonds, CO_2 has a significant quadrupole moment, and these sites encourage stronger electrostatic interactions with it. Nitrogen doping has been shown to improve hydrogen bonding between carbon surfaces and CO_2 molecules, according to FTIR measurements and quantum chemical calculations [76]. The usefulness of nitrogen doping is still debatable, nevertheless, because different research has produced contradictory findings about the functions of different nitrogen species, including pyridinic, pyridonic, and quaternary nitrogen, in CO_2 capture [77]. Similarly, the significance of oxygen-containing functional groups in CO_2 adsorption has been extensively researched but remains equivocal. Oxygen functional groups, including carboxylic acids and hydroxyl groups, are thought to improve CO_2 absorption by increasing negative surface charge and binding sites. These groups interact well with weakly acidic CO_2 molecules, especially when present in both sp^2 and sp^3 hybridised forms, increasing surface basicity and adsorption capacity. Thermal treatments at elevated pyrolysis temperatures have been shown to increase the occurrence of certain oxygen functions. However, just like nitrogen, the effect of oxygen on adsorption performance differs from a mechanism point of view [78]. Some researchers claim that oxygen-containing groups promote CO_2 adsorption, others report no significant effect or even contradictory results on selectivity toward other gases such as CH_4. These inconsistencies highlight the intricate interplay of surface chemistry, pore structure, and gas-solid interactions that control overall sorption activity. Because of its bigger atomic size and propensity to generate defects and modify the electrical configuration of carbon materials. Sulphur doping adds another dimension of surface modification. Sulphur's polar and acidic functional groups interact significantly with CO_2 [79]. CO_2 adsorption is enhanced by interactions with sulphur oxides (168.5 eV) and neutral sulphur (163.7-165.0 eV) via polar and acid-base processes. Sulphur atom's lone-pair electrons boost local surface reactivity and move the Fermi level into the conduction band, boosting electron transfer and adsorption. Strong interactions between sulphur groups and CO_2 molecules lead to considerable adsorption improvements [80]. Seema *et al.* [81] found a direct correlation between oxidised sulphur content and CO_2 capacity in sulphur-doped porous carbons. Furthermore, dual heteroatom doping techniques, such as nitrogen and sulphur co-doping, have shown synergistic effects, resulting in extraordinary CO_2 absorption. For example, N, S-doped porous carbons produced from corn starch obtained CO_2 adsorption as high as 12.03 mmol g^{-1} at 0°C and 1bar. Despite these encouraging results, the synthesis of heteroatom-doped porous carbons frequently requires expensive, toxic, and corrosive chemicals, and the methods are typically complex and time-consuming.

5.13 Hydrophobicity

In industrial contexts, CO_2 extraction from moist flue gas streams requires the use of porous materials with intrinsic hydrophobicity or even super-hydrophobicity, as well as excellent stability under humid conditions [82]. Zeolites have strong CO_2 adsorption characteristics, but their hydrophilic surface limits their efficiency in humid environments. Hydrophobicity, defined as water contact angles between 90°C and 180°C, indirectly enhances gas adsorption by rejecting water molecules from the sorbent surface and maintains pore accessibility for CO_2 molecules [83]. Surface functionalisation techniques have been used to boost both CO_2 affinity and hydrophobicity. Among these, amine functionalisation is widely employed to provide non-hydrophilic sites in the pore network, enhancing adsorption selectivity under humid conditions. However, such changes come with trade-offs. Grafting bulky hydrophobic groups reduces porosity and accessible surface area, severely impacting overall CO_2 absorption capacity.

In addition, the chemical makeup of the biomass precursor and its treatment before pyrolysis influence the hydrophobic properties of the resultant carbon material. For example, high lignin content increases hydrophobicity, whereas cellulose and hemicellulose are associated with moderate to low hydrophobicity. During pyrolysis, the progressive loss of hydrophilic functional groups as temperature increases adds to increased surface hydrophobicity and higher water contact angles [84]. In contrast, the addition of nitrogen-containing functions tends to improve wettability and hydrophilicity. Nitrogen dopants, when used as structural flaws or functional groups, at times reduce gas adsorption in damp circumstances by increasing water uptake [85]. Similarly, the presence of oxygen-containing functional groups, particularly polar ones, enhances surface hydrophilicity by forming hydrogen bonds between water molecules and oxygen atoms on the carbon surface. Carbon frameworks with functional groups such as hydroxyls (-OH) and carboxylic acids (-COOH), as well as greater O/C ratios, are more polar and hydrophilic, reducing CO_2 adsorption in humid conditions [86]. In summary, the hydrophilicity of porous carbon materials, largely governed by the presence and type of surface functional groups, acts as a barrier to efficient CO_2 adsorption under humid conditions. Therefore, a careful balance between surface polarity, functionalisation, and porosity is essential while designing sorbents for real-world gas separation applications.

5.14 Future Perspectives and Conclusion

Despite intense international efforts, there isn't yet a single, all-encompassing solution for efficient CO_2 capture. It is clear from the problem's intricacy that a variety of technological approaches will be necessary. Stability may be more crucial in some situations than selectivity, which is crucial in others. The creation of cutting-edge materials with exceptional separation performance is essential for revolutionary advancement, even while streamlining industrial operations and shrinking plant footprints can provide small, incremental gains. Beyond CO_2 sequestration, these advancements have implications for solar fuel synthesis, gas purification, and hydrogen production. Although it is outside the purview of this analysis, DAC of CO_2 is a growing field that merits attention and calls for more study and development. There is a lot of potential in hybrid materials, especially for MOFs, as discussed. For instance, adding MOFs to block copolymers that contain hydrophobic polymers may improve their water resistance. MOF-based membranes could perform better than conventional fixed-bed adsorption devices if scalable integration approaches can be created. Achieving molecular-level control over material structures is a significant scientific task. For the design and optimisation of adsorbents, this necessitates sophisticated characterisation and computational methods. To speed up discovery, high-throughput characterisation techniques ought to be used with high-throughput synthesis. Measurements of breakthrough curves and CO_2 adsorption isotherms are crucial for crystalline materials. Adsorption and permeation investigations on thin films, on the other hand, are better suitable for polymers. Important questions regarding adsorbate loading and interaction locations can be answered with the use of structural analysis and gas uptake data. Since it determines the energy needed for material regeneration, figuring out the enthalpy of adsorption is crucial. The comprehension of molecular movement is similarly important. Important topics include where and how CO_2 attaches, how structures change during gas adsorption, and how the presence of various affects adsorption behaviour, which needs further exploration. Methods like solid-state NMR and resonant soft X-ray spectroscopy are essential for connecting molecular structure with diffusion behaviour. In the end, establishing a correlation between microscopic mobility and macroscopic diffusion will yield important information about selective gas transport. Computational modelling is necessary in addition to experimental techniques. Thousands of materials can be virtually screened with the help of molecular simulations that can forecast adsorption and transport characteristics. Such modelling is very simple for crystalline structures. However, the complexity rises dramatically for membranes made of amorphous polymers. Accurate material performance forecasts need the development of trustworthy force fields and structural models. An interdisciplinary strategy integrating material synthesis, sophisticated characterisation, and predictive modelling is needed to make significant advances.

Furthermore, it is crucial to assess how promising materials will perform in actual separation procedures. Economic assessments must take into account the life-cycle costs of capture, regeneration, and sequestration in addition to laboratory viability. The industrial feasibility of materials, especially membranes and MOFs, depends on their synthesis cost and scalability.

Even if a thorough techno-economic analysis is outside the purview of this book chapter, it is an important topic for further research (Insights are provided in the last chapter). Several component technologies, such as pipeline transport and CO_2 compression, have already been developed, even though fully integrated commercial CCS systems are not yet operational. Capture material testing on a pilot scale is progressing well, but there is an urgent need for more rapid development and implementation of affordable materials. With large pilot programmes concentrating on ocean injection, mineral carbonation, and geological sequestration—such as saline aquifers and depleted oil/gas reservoirs—long-term CO_2 storage still needs exploration. The urgency of climate change necessitates swift and consistent action despite the numerous scientific, engineering, regulatory, and financial obstacles. We are in a great position as a scientific community to spearhead the development of next-generation CO_2 capture technology and to create workable solutions for reducing carbon emissions worldwide. At the end of this chapter, light is shed on the surface related aspects of the materials employed for the CCS technology.

REFERENCES

[1] D.M. D'Alessandro, B. Smit, J.R. Long, Carbon dioxide capture: Prospects for new materials, Angew. Chemie - Int. Ed. 49 (2010) 6058–6082. https://doi.org/10.1002/anie.201000431.

[2] Intergovernmental Panel on Climate Change, IPCC Special Report on Carbon Dioxide Capture and Storage, 2011.

[3] Madhusudhana M. Devadiga et al., Describing critical barriers in commercial deployment of CO2 capture and utilization technologies, Energy Environ. Sci., 2026,19, 743-790https://doi.org/10.1039/D6EE00134C

[4] R. Stuart Haszeldine, Carbon capture and storage: how green can black be?, Science (80). 325 (2009) 1647–1652. https://doi.org/10.1126/science.1172246.

[5] K.W. Bandilla, Carbon capture and storage, in: Futur. Energy Improv. Sustain. Clean Options Our Planet, 2020: pp. 669–692. https://doi.org/10.1016/B978-0-08-102886-5.00031-1.

[6] G. Göttlicher, R.P.E.C. and Management, undefined 1997, Comparison of CO_2 removal systems for fossil-fuelled power plant processesQ1Energy Conversion and Management; H-Index: 210SJR: Q1 CORE: NA ABDC: NA FT50: NA 2Energy Conversion and Management; H-Index: 210VHB: NA FNEGE: NA CoNRS: NA HCERE: NA CCF: NA BFI: 2 AJG: NA +, ElsevierG Göttlicher, R PruschekEnergy Convers. Manag. 1997•Elsevier (n.d.). https://www.sciencedirect.com/science/article/pii/S0196890496002658.

[7] S. Castro-Pardo, S. Bhattacharyya, R.M. Yadav, A.P. de Carvalho Teixeira, M.A. Campos Mata, T. Prasankumar, M.A. Kabbani, M.G. Kibria, T. Xu, S. Roy, P.M. Ajayan, A comprehensive overview of carbon dioxide capture: From materials, methods to industrial status, Mater. Today 60 (2022) 227–270. https://doi.org/10.1016/J.MATTOD.2022.08.018.

[8] S. Choi, J.H. Drese, C.W. Jones, Adsorbent materials for carbon dioxide capture from large anthropogenic point sources, ChemSusChem 2 (2009) 796–854. https://doi.org/10.1002/cssc.200900036.

[9] G. Singh, J. Lee, A. Karakoti, R. Bahadur, J. Yi, D. Zhao, K. Albahily, A. Vinu, Emerging trends in porous materials for CO2capture and conversion, Chem. Soc. Rev. 49 (2020) 4360–4404. https://doi.org/10.1039/d0cs00075b.

[10] S.D. Kenarsari, D. Yang, G. Jiang, S. Zhang, J. Wang, A.G. Russell, Q. Wei, M. Fan, Review of recent advances in carbon dioxide separation and capture, RSC Adv. 3 (2013) 22739–22773. https://doi.org/10.1039/c3ra43965h.

[11] A.A. Olajire, CO_2 capture and separation technologies for end-of-pipe applications – A review, Energy 35 (2010) 2610–2628. https://doi.org/10.1016/J.ENERGY.2010.02.030.

[12] S. Kumar, R. Srivastava, J. Koh, Utilization of zeolites as CO_2 capturing agents: Advances and future perspectives, J. CO2 Util. 41 (2020) 101251. https://doi.org/10.1016/J.JCOU.2020.101251.

[13] Y. Liu, J. Shi, J. Chen, Q. Ye, H. Pan, Z. Shao, Y. Shi, Dynamic performance of CO_2 adsorption with tetraethylenepentamine-loaded KIT-6, Microporous Mesoporous Mater. 134 (2010) 16–21. https://doi.org/10.1016/J.MICROMESO.2010.05.002.

[14] R. Serna-Guerrero, A. Sayari, Modeling adsorption of CO_2 on amine-functionalized mesoporous silica. 2: Kinetics and breakthrough curves, Chem. Eng. J. 161 (2010) 182–190. https://doi.org/10.1016/J.CEJ.2010.04.042.

[15] R. Serna-Guerrero, Y. Belmabkhout, A. Sayari, Influence of regeneration conditions on the cyclic performance of amine-grafted mesoporous silica for CO_2 capture: An experimental and statistical study, Chem. Eng. Sci. 65 (2010) 4166–4172. https://doi.org/10.1016/J.CES.2010.04.029.

[16] M.G. Plaza, C. Pevida, A. Arenillas, F. Rubiera, J.J. Pis, CO_2 capture by adsorption with nitrogen enriched carbons, Fuel 86 (2007) 2204–2212. https://doi.org/10.1016/J.FUEL.2007.06.001.

[17] P. Cabrera-Sanfelix, Adsorption and Reactivity of CO_2 on Defective Graphene Sheets, J. Phys. Chem. A 113 (2008) 493–498. https://doi.org/10.1021/JP807087Y.

[18] A.B. Bourlinos, T.A. Steriotis, M. Karakassides, Y. Sanakis, V. Tzitzios, C. Trapalis, E. Kouvelos, A. Stubos, Synthesis, characterization and gas sorption properties of a molecularly-derived graphite oxide-like foam, Carbon N. Y. 45 (2007) 852–857. https://doi.org/10.1016/J.CARBON.2006.11.008.

[19] S.N. Talapaneni, G. Singh, I.Y. Kim, K. AlBahily, A.H. Al-Muhtaseb, A.S. Karakoti, E. Tavakkoli, A. Vinu, Nanostructured Carbon Nitrides for CO_2 Capture and Conversion, Adv. Mater. 32 (2020) 1904635. https://doi.org/10.1002/ADMA.201904635.

[20] G. Xu, H. Zhang, J. Wei, H.X. Zhang, X. Wu, Y. Li, C. Li, J. Zhang, J. Ye, Integrating the g-C_3N_4 Nanosheet with B–H Bonding Decorated Metal–Organic Framework for CO_2 Activation and Photoreduction, ACS Nano 12 (2018) 5333–5340. https://doi.org/10.1021/ACSNANO.8B00110.

[21] H.L. Peng, F.Y. Zhong, J.B. Zhang, J.Y. Zhang, P.K. Wu, K. Huang, J.P. Fan, L.L. Jiang, Graphitic Carbon Nitride Functionalized with Polyethylenimine for Highly Effective Capture of Carbon Dioxide, Ind. Eng. Chem. Res. 57 (2018) 11031–11038. https://doi.org/10.1021/ACS.IECR.8B02275.

[22] J. Xu, K. Shen, B. Xue, Y.X. Li, Microporous carbon nitride as an effective solid base catalyst for Knoevenagel condensation reactions, J. Mol. Catal. A Chem. 372 (2013) 105–113. https://doi.org/10.1016/J.MOLCATA.2013.02.019.

[23] L. Chen, J. Song, Tailored Graphitic Carbon Nitride Nanostructures: Synthesis, Modification, and Sensing Applications, Adv. Funct. Mater. 27 (2017) 1702695. https://doi.org/10.1002/ADFM.201702695.

[24] X. Tan, L. Kou, H.A. Tahini, S.C. Smith, Conductive Graphitic Carbon Nitride as an Ideal Material for Electrocatalytically Switchable CO_2 Capture, Sci. Rep. 5 (2015) 1–8. https://doi.org/10.1038/SREP17636;SUBJMETA.

[25] K.S. Lakhi, W.S. Cha, S. Joseph, B.J. Wood, S.S. Aldeyab, G. Lawrence, J.H. Choy, A. Vinu, Cage type mesoporous carbon nitride with large mesopores for CO_2 capture, Catal. Today 243 (2015) 209–217. https://doi.org/10.1016/J.CATTOD.2014.08.036.

[26] F.E.C. Othman, N. Yusof, S. Samitsu, N. Abdullah, M.F. Hamid, K. Nagai, M.N.Z. Abidin, M.A. Azali, A.F. Ismail, J. Jaafar, F. Aziz, W.N.W. Salleh, Activated carbon nanofibers incorporated metal oxides for CO_2 adsorption: Effects of different type of metal oxides, J. CO_2 Util. 45 (2021) 101434. https://doi.org/10.1016/J.JCOU.2021.101434.

[27] Z. Liu, W. Li, P.Z. Moghadam, S. Li, Screening adsorbent-water adsorption heat pumps based on an experimental water adsorption isotherm database, Sustain. Energy Fuels 5 (2021) 1075–1084. https://doi.org/10.1039/d0se01824d.

[28] S. Fu, Q. Fang, A. Li, Z. Li, J. Han, X. Dang, W. Han, Accurate characterization of full pore size distribution of tight sandstones by low-temperature nitrogen gas adsorption and high-pressure mercury intrusion combination method, Energy Sci. Eng. 9 (2021) 80–100. https://doi.org/10.1002/ESE3.817.

[29] R. M., K.G. Ayappa, Influence of the extent of hydrophobicity on water organization and dynamics on 2D graphene oxide surfaces, Phys. Chem. Chem. Phys. 24 (2022) 14909–14923. https://doi.org/10.1039/d1cp03962h.

[30] A. Wang, K. Sun, R. Xu, Y. Sun, J. Jiang, Cleanly synthesizing rotten potato-based activated carbon for supercapacitor by self-catalytic activation, J. Clean. Prod. 283 (2021) 125385. https://doi.org/10.1016/J.JCLEPRO.2020.125385.

[31] P. Ozpinar, C. Dogan, H. Demiral, U. Morali, S. Erol, C. Samdan, D. Yildiz, I. Demiral, Activated carbons prepared from hazelnut shell waste by phosphoric acid activation for supercapacitor electrode applications and comprehensive electrochemical analysis, Renew. Energy 189 (2022) 535–548. https://doi.org/10.1016/J.RENENE.2022.02.126.

[32] Y.; Wang, J.; Wang, X.; Zhang, D.; Bhattacharyya, E.M. Sabolsky, Y. Wang, J. Wang, X. Zhang, D. Bhattacharyya, E.M. Sabolsky, Quantifying Environmental and Economic Impacts of Highly Porous Activated Carbon from Lignocellulosic Biomass for High-Performance Supercapacitors, Energies 2022, Vol. 15, Page 351 15 (2022) 351. https://doi.org/10.3390/EN15010351.

[33] S.J. Allen, G. Mckay, J.F. Porter, Adsorption isotherm models for basic dye adsorption by peat in single and binary component systems, J. Colloid Interface Sci. 280 (2004) 322–333. https://doi.org/10.1016/J.JCIS.2004.08.078.

[34] G. Limousin, J.P. Gaudet, L. Charlet, S. Szenknect, V. Barthès, M. Krimissa, Sorption isotherms: A review on physical bases, modeling and measurement, Appl. Geochemistry 22 (2007) 249–275. https://doi.org/10.1016/J.APGEOCHEM.2006.09.010.

[35] M. Ghiaci, A. Abbaspur, R. Kia, F. Seyedeyn-Azad, Equilibrium isotherm studies for the sorption of benzene, toluene, and phenol onto organo-zeolites and as-synthesized MCM-41, Sep. Purif. Technol. 40 (2004) 217–229. https://doi.org/10.1016/J.SEPPUR.2004.03.001.

[36] M.C. Ncibi, Applicability of some statistical tools to predict optimum adsorption isotherm after linear and non-linear regression analysis, J. Hazard. Mater. 153 (2008) 207–212. https://doi.org/10.1016/J.JHAZMAT.2007.08.038.

[37] A.A. Alhwaige, Novel biobased chitosan/polybenzoxazine cross-linked polymers and advanced carbon aerogels for CO_2 adsorption, (2014).

[38] V.K. Singh, E. Anil Kumar, Measurement and analysis of adsorption isotherms of CO_2 on activated carbon, Appl. Therm. Eng. 97 (2016) 77–86. https://doi.org/10.1016/j.applthermaleng.2015.10.052.

[39] M.G. Lee, Y. Kim, Enhanced antibiotic removal by waste coffee grounds prepared via water washing and KOH activation, Korean J. Chem. Eng. 40 (2023) 2489–2496. https://doi.org/10.1007/S11814-023-1550-8/METRICS.

[40] G.K. Parshetti, S. Chowdhury, R. Balasubramanian, Biomass derived low-cost microporous adsorbents for efficient CO2 capture, Fuel 148 (2015) 246–254. https://doi.org/10.1016/J.FUEL.2015.01.032.

[41] M.S. Khosrowshahi, H. Mashhadimoslem, H. Shayesteh, G. Singh, E. Khakpour, X. Guan, M. Rahimi, F. Maleki, P. Kumar, A. Vinu, Natural Products Derived Porous Carbons for CO_2 Capture, Adv. Sci. 10 (2023) 2304289. https://doi.org/10.1002/ADVS.202304289.

[42] K. Vijayaraghavan, T.V.N. Padmesh, K. Palanivelu, M. Velan, Biosorption of nickel(II) ions onto Sargassum wightii: Application of two-parameter and three-parameter isotherm models, J. Hazard. Mater. 133 (2006) 304–308. https://doi.org/10.1016/J.JHAZMAT.2005.10.016.

[43] F.H.M. F., Over the adsorption in solution., J. Phys. Chem. 57 (1906) 1100–1107.

[44] I.L. The constitution and fundamental properties of solids and liquids. Part I. Solids.Q1Journal of the American chemical society, ACS Publ. 38 (1916) 2221–2295. https://doi.org/10.1021/JA02268A002.

[45] V. Tempkin, M.I. Pyzhev, Kinetics of ammonia synthesis on promoted iron catalyst, Acta Phys. Chim. USSR 12 (1940) 327–356.

[46] F. Stoeckli, Dubinin's theory and its contribution to adsorption science, Russ. Chem. Bull. 50 (2001) 2265–2272. https://doi.org/10.1023/A:1015054922994/METRICS.

[47] M.M. and R. Dubinin, Equation of the Characteristic Curve of Activated Charcoal Proceedings of the Academy of Sciences. Scientific Research Publishing, Phys. Chem. Sect. USSR 55 (1947) 331–333. https://www.scirp.org/reference/referencespapers?referenceid=1387660.

[48] G. Halsey, Physical Adsorption on Non-Uniform Surfaces, J. Chem. Phys. 16 (1948) 931–937. https://doi.org/10.1063/1.1746689.

[49] N. Ayawei, A.N. Ebelegi, D. Wankasi, Modelling and Interpretation of Adsorption Isotherms, J. Chem. 2017 (2017) 3039817. https://doi.org/10.1155/2017/3039817.

[50] F. Brouers, F. Brouers, Statistical Foundation of Empirical Isotherms, Open J. Stat. 4 (2014) 687–701. https://doi.org/10.4236/OJS.2014.49064.

[51] W. McMillan, E.T.T.J. of P. Chemistry, undefined 1951, The assumptions of the BET theory.NANo ranking found for "The Journal of Physical Chemistry, ACS Publ. 55 (1951) 17–20. https://doi.org/10.1021/J150484A003.

[52] R. Sips, On the Structure of a Catalyst Surface, J. Chem. Phys. 16 (1948) 490–495. https://doi.org/10.1063/1.1746922.

[53] M. Temkin, V. Pyzhev, Kinetics of the synthesis of ammonia on promoted iron catalysts, Jour. Phys. Chem. (U.S.S.R.) 13 (1939) 851–867.

[54] H. Wang, H. Wang, G. Liu, Q. Yan, In-situ pyrolysis of Taihu blue algae biomass as appealing porous carbon adsorbent for CO_2 capture: Role of the intrinsic N, Sci. Total Environ. 771 (2021) 145424. https://doi.org/10.1016/J.SCITOTENV.2021.145424.

[55] X. Xu, Z. Xu, B. Gao, L. Zhao, Y. Zheng, J. Huang, D.C.W. Tsang, Y.S. Ok, X. Cao, New insights into CO_2 sorption on biochar/Fe oxyhydroxide composites: Kinetics, mechanisms, and in situ characterization, Chem. Eng. J. 384 (2020) 123289. https://doi.org/10.1016/J.CEJ.2019.123289.

[56] W. Sghaier, Y. Ben Torkia, M. Bouzid, A. Ben Lamine, CO_2 adsorption investigation by statistical physics: Thermodynamic analysis for cooling cycle application, J. Environ. Chem. Eng. 9 (2021) 105108. https://doi.org/10.1016/J.JECE.2021.105108.

[57] H. Marsh, B. Rand, The characterization of microporous carbons by means of the dubinin-radushkevich equation, J. Colloid Interface Sci. 33 (1970) 101–116. https://doi.org/10.1016/0021-9797(70)90077-9.

[58] L. Li, J. Yang, J. Li, J. Li, Y. Chen, Separation of CO_2/CH_4 and CH_4/N_2 mixtures by M/DOBDC: A detailed dynamic comparison with MIL-100(Cr) and activated carbon, Microporous Mesoporous Mater. 198 (2014) 236–246. https://doi.org/10.1016/J.MICROMESO.2014.07.041.

[59] L. Lei et al, Taming structure and modulating carbon dioxide (CO_2) adsorption isosteric heat of nickel-based metal organic framework (MOF-74 (Ni)) for remarkable CO_2 Capture, Journal Colloid Interface Sci. 2022, Pages 132-145 Elsevier. https://www.sciencedirect.com/science/article/pii/S0021979721023079.

[60] H. Zhou, C. Rayer, A.R. Antonangelo, N. Hawkins, M. Carta, Adjustable Functionalization of Hyper-Cross-Linked Polymers of Intrinsic Microporosity for Enhanced CO_2 Adsorption and Selectivity over N_2 and CH_4, ACS Appl. Mater. Interfaces 14 (2022) 20997–21006. https://doi.org/10.1021/ACSAMI.2C02604.

[61] D. Xu, P. Xiao, J. Zhang, G. Li, G. Xiao, P.A. Webley, Y. Zhai, Effects of water vapour on CO_2 capture with vacuum swing adsorption using activated carbon, Chem. Eng. J. 230 (2013) 64–72. https://doi.org/10.1016/J.CEJ.2013.06.080.

[62] C.A. Grande, A.E. Rodrigues, Electric Swing Adsorption for CO_2 removal from flue gases, Int. J. Greenh. Gas Control 2 (2008) 194–202. https://doi.org/10.1016/S1750-5836(07)00116-8.

[63] C. Shen, Z. Liu, P. Li, J. Yu, Two-Stage VPSA Process for CO_2 Capture from Flue Gas Using Activated Carbon Beads, Ind. Eng. Chem. Res. 51 (2012) 5011–5021. https://doi.org/10.1021/IE202097Y.

[64] A.K. Rajagopalan, A.M. Avila, A. Rajendran, Do adsorbent screening metrics predict process performance? A process optimisation based study for post-combustion capture of CO_2, Int. J. Greenh. Gas Control 46 (2016) 76–85. https://doi.org/10.1016/j.ijggc.2015.12.033.

[65] J. Sreńscek-Nazzal, U. Narkiewicz, A.W. Morawski, R.J. Wróbel, B. Michalkiewicz, Comparison of Optimized Isotherm Models and Error Functions for Carbon Dioxide Adsorption on Activated Carbon, J. Chem. Eng. Data 60 (2015) 3148–3158. https://doi.org/10.1021/ACS.JCED.5B00294.

[66] F. Raganati, F. Miccio, P. Ammendola, Adsorption of Carbon Dioxide for Post-combustion Capture: A Review, Energy and Fuels 35 (2021) 12845–12868. https://doi.org/10.1021/acs.energyfuels.1c01618.

[67] B. Chen, Z. Yang, G. Ma, D. Kong, W. Xiong, J. Wang, Y. Zhu, Y. Xia, Heteroatom-doped porous carbons with enhanced carbon dioxide uptake and excellent methylene blue adsorption capacities, Microporous Mesoporous Mater. 257 (2018) 1–8. https://doi.org/10.1016/J.MICROMESO.2017.08.026.

[68] D. Saha, M.J. Kienbaum, Role of oxygen, nitrogen and sulfur functionalities on the surface of nanoporous carbons in CO_2 adsorption: A critical review, Microporous Mesoporous Mater. 287 (2019) 29–55. https://doi.org/10.1016/J.MICROMESO.2019.05.051

[69] S. Dutta, A. Bhaumik, K.C.W. Wu, Hierarchically porous carbon derived from polymers and biomass: effect of interconnected pores on energy applications, Energy Environ. Sci. 7 (2014) 3574–3592. https://doi.org/10.1039/C4EE01075B.

[70] S. Liu, A. Busch, J. Ma, S. Sang, T. Wang, Y. Du, H. Fang, The effects of supercritical CO_2 on mesopore and macropore structure in bituminous and anthracite coal, Fuel 223 (2018) 32–43. https://doi.org/10.1016/J.FUEL.2018.03.036.

[71] M. Vorokhta, J. Morávková, M. Dopita, A. Zhigunov, M. Šlouf, R. Pilař, P. Sazama, Effect of micropores on CO_2 capture in ordered mesoporous CMK-3 carbon at atmospheric pressure, Adsorption 27 (2021) 1221–1236. https://doi.org/10.1007/S10450-021-00322-Y/FIGURES/13.

[72] E. Maruccia, M.A.O. Lourenço, T. Priamushko, M. Bartoli, S. Bocchini, F.C. Pirri, G. Saracco, F. Kleitz, C. Gerbaldi, Nanocast nitrogen-containing ordered mesoporous carbons from glucosamine for selective CO_2 capture, Mater. Today Sustain. 17 (2022) 100089. https://doi.org/10.1016/J.MTSUST.2021.100089.

[73] J. Serafin, B. Dziejarski, O.F. Cruz Junior, J. Sreńscek-Nazzal, Design of highly microporous activated carbons based on walnut shell biomass for H_2 and CO_2 storage, Carbon N. Y. 201 (2023) 633–647. https://doi.org/10.1016/J.CARBON.2022.09.013.

[74] H. Wei, S. Deng, B. Hu, Z. Chen, B. Wang, J. Huang, G. Yu, Granular Bamboo-Derived Activated Carbon for High CO_2 Adsorption: The Dominant Role of Narrow Micropores, ChemSusChem 5 (2012) 2354–2360. https://doi.org/10.1002/CSSC.201200570.

[75] A. Vinu, Two-Dimensional Hexagonally-Ordered Mesoporous Carbon Nitrides with Tunable Pore Diameter, Surface Area and Nitrogen Content, Adv. Funct. Mater. 18 (2008) 816–827. https://doi.org/10.1002/ADFM.200700783.

[76] M. Zhu, W. Cai, F. Verpoort, J. Zhou, Preparation of pineapple waste-derived porous carbons with enhanced CO_2 capture performance by hydrothermal carbonation-alkali metal oxalates assisted thermal activation process, Chem. Eng. Res. Des. 146 (2019) 130–140. https://doi.org/10.1016/J.CHERD.2019.03.044.

[77] M. Sevilla, J.B. Parra, A.B. Fuertes, Assessment of the Role of Micropore Size and N-Doping in CO_2 Capture by Porous Carbons, ACS Appl. Mater. Interfaces 5 (2013) 6360–6368. https://doi.org/10.1021/AM401423B.

[78] X. Song, L. Wang, J. Gong, X. Zhan, Y. Zeng, Exploring a New Method to Study the Effects of Surface Functional Groups on Adsorption of CO_2 and CH_4 on Activated Carbons, Langmuir 36 (2020) 3862–3870. https://doi.org/10.1021/ACS.LANGMUIR.9B03475.

[79] S. Hao, J. Wen, X. Yu, W. Chu, Effect of the surface oxygen groups on methane adsorption on coals, Appl. Surf. Sci. 264 (2013) 433–442. https://doi.org/10.1016/J.APSUSC.2012.10.040.

[80] R.R. Da Silva, J.H.S. Torres, Y. Kopelevich, Indication of Superconductivity at 35 K in Graphite-Sulfur Composites, Phys. Rev. Lett. 87 (2001) 147001. https://doi.org/10.1103/PhysRevLett.87.147001.

[81] H. Seema, K.C. Kemp, N.H. Le, S.W. Park, V. Chandra, J.W. Lee, K.S. Kim, Highly selective CO_2 capture by S-doped microporous carbon materials, Carbon N. Y. 66 (2014) 320–326. https://doi.org/10.1016/J.CARBON.2013.09.006.

[82] P. Yang, L. Rao, W. Zhu, L. Wang, R. Ma, F. Chen, G. Lin, X. Hu, Porous Carbons Derived from Sustainable Biomass via a Facile One-Step Synthesis Strategy as Efficient CO_2 Adsorbents, Ind. Eng. Chem. Res. 59 (2020) 6194–6201. https://doi.org/10.1021/ACS.IECR.0C00073.

[83] M. Lee, S. Hong, D. Kim, E. Kim, K. Lim, J.C. Jung, H. Richter, J.H. Moon, N. Choi, J. Nam, J. Choi, Chabazite-Type Zeolite Membranes for Effective CO_2 Separation: The Role of Hydrophobicity and Defect Structure, ACS Appl. Mater. Interfaces 11 (2019) 3946–3960. https://doi.org/10.1021/ACSAMI.8B18854.

[84] S. V. Vassilev, C.G. Vassileva, V.S. Vassilev, Advantages and disadvantages of composition and properties of biomass in comparison with coal: An overview, Fuel 158 (2015) 330–350. https://doi.org/10.1016/J.FUEL.2015.05.050.

[85] B.M. Matsagar, R.X. Yang, S. Dutta, Y.S. Ok, K.C.W. Wu, Recent progress in the development of biomass-derived nitrogen-doped porous carbon, J. Mater. Chem. A 9 (2021) 3703–3728. https://doi.org/10.1039/d0ta09706c.

[86] H. Banna Motejadded Emrooz, M. Maleki, M. Shokouhimehr, Excellent adsorption of orange acid II on a water fern–derived micro- and mesoporous carbon, J. Taiwan Inst. Chem. Eng. 102 (2019) 99–109. https://doi.org/10.1016/J.JTICE.2019.05.009.

Chapter 6

TRANSLATING ADVANCED MATERIAL INTERFACES INTO PRACTICAL CO_2 CAPTURE

Abstract: This chapter provides an overview of carbon capture technology, tracing its journey from laboratory research to industrial-scale applications. It highlights the critical gap that exists between the efficiency of materials demonstrated in the lab and their real-world performance in industrial systems. The main challenge underlying this gap, "the so-called valley of death," stems from limitations in understanding, engineering design, and material interface advancements, all of which span multiple domains from the nano- to micro-scale. Following a brief introduction, the chapter presents an in-depth discussion on the foundational principles of scaling up carbon capture processes. This includes the transition from gram-scale to tonne-scale synthesis, emphasising the role of engineering strategies and precursor selection. The discussion further extends to different classes of materials: sorbents, MOFs, and membrane technologies, analysing their strengths and shortcomings. Attention is also given to the integration of these materials into real operating conditions, focusing on the need to harmonise material properties with the thermodynamic and kinetic requirements of the capture process. Challenges such as industrial retrofitting and system-level obstacles are discussed to illustrate the complexity of bridging laboratory success with industrial deployments. Finally, the chapter outlines strategies for overcoming these barriers, moving from research focused on isolated material properties to a more holistic, system-level approach. Emphasis is placed on designing integrated strategies that balance cost, durability, and performance. Such system-oriented methodologies are presented as a pathway towards advancing the next generation of carbon capture technologies, ensuring their economic viability and large-scale applicability.

Keywords: Carbon Capture, Material Interfaces, Scale-Up, Nanocomposites, MOFs, Membranes, Techno-Economic Analysis.

6.1 Introduction

Globally, significant initiatives are underway to mitigate the impacts of climate change, which, if left unaddressed, could lead to catastrophic crises. Among the various strategies, the reduction of atmospheric CO_2 concentration through CCUS has emerged as one of the most promising approaches [1]. This method has been strongly underscored and recognised by the Intergovernmental Panel on Climate Change (IPCC). It is widely endorsed by governmental and private agencies across the world as a key pathway to achieving the net-zero emission goal [2, 3]. Despite its global significance, the sobering reality is that only a small fraction of these technologies has been successfully implemented in real-world practice. Over the past decade, a wide range of advanced materials has been developed that exhibit exceptional performance in laboratory studies. However, the majority of these materials fail to deliver when scaled from controlled laboratory conditions to industrial-scale applications. This critical challenge, often referred to as the "valley of death," remains one of the greatest barriers to the successful deployment of next-generation materials for carbon capture [4, 5]. The transition from producing grams of adsorbents at the laboratory scale to manufacturing tonnes for industrial applications requires not only meticulous engineering but also a thorough consideration of scientific, technological, and economic factors. Unfortunately, many of these aspects are overlooked during the early phases of material development, creating major hurdles in scaling up and realising the full potential of CCUS technologies.

Among the many challenges faced at both the fundamental and advanced levels, the core problem in scaling carbon capture technologies lies in understanding the material interface. This interface is a multidimensional concept, far more complex than a simple particle surface [6]. The success of transitioning from laboratory research to industrial application depends on mastering this interface through a deep scientific understanding, followed by careful engineering design. At its most basic level, the interface represents the active site where CO_2 molecules attach, either through physical adsorption or chemical bonding. Yet, its role extends beyond this microscopic interaction. For instance, in MOFs, the interface governs how the MOF particles interact with binders to form mechanically stable pellets suitable for use in packed beds or reactors. In mixed-matrix systems, it is the point of contact where the filler must integrate seamlessly with the polymer matrix to avoid defects that could compromise performance [7]. When scaled up to industrial applications, the concept of the interface expands to engineered forms such as monoliths, fibres, or films. These structures must withstand harsh conditions, including high temperatures, corrosive contaminants, and significant mechanical stresses. Thus, the interface plays a critical role not only in ensuring performance but also in determining the durability and economic viability of the material [6]. In this hierarchical manner, from the nanoscale to the macroscale, the interface serves as the linchpin of material performance. Properly designing and engineering it will offer the most promising route in bridging the gap between laboratory innovations and real-world industrial applications.

Looking closely at the origins of carbon capture research, one recurring theme becomes evident: interfacial failures that arise when laboratory-developed materials are scaled up for practical applications[7]. These failures vary across different classes of materials but consistently emerge as a critical bottleneck.

For instance, in the case of amine-functionalised sorbents, the key lies in the stability of the interface between the organic amine groups and their porous supports. While these materials exhibit excellent CO_2 adsorption under controlled conditions, their performance deteriorates significantly under flue gas environments containing steam and other gases. This exposure initiates amine leaching at the interface, ultimately reducing both efficiency and durability. Similarly, MOFs demonstrate outstanding CO_2 capture capacity in their powdered form [8, 9]. However, at the industrial scale, they encounter severe interfacial challenges, most notably during pelletisation, where maintaining pore accessibility is difficult, thereby undermining their effectiveness. In the case of membranes, interfacial issues are equally pressing. The goal of achieving uniform filler distribution during fabrication is often hindered by poor adhesion at the filler-polymer interface. This weak bonding creates defects and non-selective voids, allowing gases to bypass the intended separation pathways and drastically reducing efficiency[10].

Thus, achieving success in real-life applications requires bridging the existing gap through a paradigm shift in research and development strategies. Rather than focusing solely on the intrinsic properties of individual materials, a comprehensive, system-level approach centred on interfacial engineering and advanced integrations must be adopted [6]. Importantly, materials should be evaluated under realistic operating conditions rather than ideal laboratory atmospheres. This includes testing in the presence of steam, assessing stability against flue gas contaminants, and verifying mechanical robustness to withstand the stresses of industrial reactors. At the same time, these efforts must remain closely tied to economic viability [6, 9]. Costs associated with raw materials, synthesis routes, energy requirements, and regeneration cycles must be carefully managed; otherwise, the resulting materials, despite their performance, will remain commercially unfeasible. Therefore, techno-economic analyses and life-cycle assessments should be prioritised from the outset. These tools will guide progress toward solutions that are not only effective but also scalable, affordable, and sustainable [5].

This chapter aims to provide a comprehensive overview of the journey of carbon capture technologies, tracing their progress from early discovery to the current state of development, while also outlining a roadmap for the future. It will begin with a discussion of the foundational principles of scaling different materials, highlighting the scientific and engineering challenges involved in synthesis, fabrication, and interface design for real-world applications, along with strategies to address these obstacles. To deepen the understanding, the chapter will examine both the successes and limitations encountered with key classes of materials, including nanocomposites, MOFs, and membranes. Special attention will be given to the role of interfaces, such as the interactions between materials and binders, fillers and polymers, or sorbents and their supports, that critically influence the practical applicability and durability of these systems. Moving forward, an in-depth discussion will focus on the challenges of system integration, emphasising the need to harmonise material properties with the operational dynamics of industrial processes. Finally, the chapter will provide a forward-looking perspective on the persistent challenges and emerging frontiers that could accelerate the development of next-generation carbon capture materials, with improved synthesis routes, advanced fabrication techniques, and enhanced performance.

6.2 Foundation Principles of Scaling-Up Advanced Materials for Carbon Capture

The advancement of materials from laboratory development to industrial-scale applications for carbon capture is both a remarkable and challenging task, guided by a set of stringent foundational principles. While the intrinsic properties of a material, such as CO_2 affinity, selectivity, and adsorption capacity, are essential drivers for CCUS applications, they alone are not sufficient to establish industrial viability [11]. For real-world deployment, materials must also be scalable in production, precisely engineered in morphology and pore architecture, and stable under diverse operating conditions. Meeting these conditions requires not only innovation in design but also a paradigm shift that translates laboratory-scale strategies into industrial-scale solutions, without compromising performance, cost-effectiveness, or durability. Although significant progress has been made, the absence of coherent strategies and oversight of these critical aspects has led many promising materials to fall short in practice. To understand why, let us explore these factors step by step [12].

The first cornerstone is material synthesis. Producing small amounts of material in the laboratory, such as MOFs or nanocomposites through solvothermal routes, is achievable with precision control of temperature, precursor choice, stirring, and purification. However, scaling the same principles to manufacture tonnes of material for a power plant is both economically and logistically unrealistic [12]. Batch-wise synthesis cannot meet industrial demand, highlighting the need to transition towards continuous or semi-continuous processes. Flow reactors, if carefully engineered, can provide better heat and mass transfer, reduce raw material usage, ensure consistency in material quality, and deliver higher throughput compared to batch systems. The main challenge lies in adapting sensitive laboratory processes for continuous operation over extended durations, which requires careful optimisation of reaction kinetics, concentrations, and cost-effectiveness. Hence, scalable synthesis depends not just on quantity but also on the ability to preserve laboratory-level properties at the industrial scale, demanding advanced analytical methods and process controls [13].

Closely tied to synthesis is material fabrication. While raw powders may demonstrate excellent CO_2 adsorption in the lab, they are often unsuitable for industrial use. In packed beds, powders can lead to unacceptably high pressure drops and risk entrainment losses in fluidised systems. Therefore, shaping materials into pellets, beads, sheets, or monoliths is essential. However, fabrication introduces another complication: the use of binders, plasticisers, or additives for mechanical strength often blocks active pores, reducing adsorption efficiency. For instance, pelletisation may require 10–30% inactive additives, and extrusion can damage pore structures. Overcoming these drawbacks requires strategies such as using sacrificial porogens to preserve porosity, designing hierarchical structures with macro- and micropores for efficient gas transport, or employing binders with inherent affinity towards CO_2. Ultimately, the fabricated form, not the powder, determines the real performance of an adsorbent [14].

Once scalable production is achieved, the material must prove its robustness in real operating environments. Flue gas streams from power plants contain CO_2, N_2, water vapour, and corrosive contaminants like O_2, SO_2, and NO_2 at elevated temperatures (40–150°C). Such conditions severely test material stability. MOFs, despite their excellent properties, can undergo irreversible collapse in humid flue gas. At the same time, amine-based sorbents may degrade in the presence of acidic gases, forming heat-stable salts with costly regeneration requirements. Thus, materials must be engineered for resilience through hydrophobic designs, oxidation-resistant functional groups, or protective core-shell architectures [15]. At the same time, mechanical integrity is critical since continuous operation in reactors subjects materials to attrition and crushing. Only materials with excellent strength, attrition resistance, and stability can sustain long-term industrial use [11].

Underlying all these factors are two essential frameworks: techno-economic analysis (TEA) and life-cycle assessment (LCA). Together, they ensure that technical success aligns with economic and environmental sustainability. TEA enables early-stage cost evaluation and optimisation, helping prioritise materials with a favourable balance between adsorption performance and regeneration cost. LCA, on the other hand, evaluates the broader environmental footprint, ensuring that the synthesis of advanced nanomaterials does not introduce hazardous by-products or excessive emissions [16]. As TEA and LCA are deeply interlinked, improvements in one may negatively affect the other; systematic, multidisciplinary collaboration is required. Scientists, engineers, and economists must work together to navigate these trade-offs and bridge the notorious "valley of death" between laboratory discovery and industrial deployment [16, 17].

6.3 Practical Demonstration for Advanced Material Interfaces in Action

The true success of a material designed for carbon capture is not defined by its laboratory characteristics alone, but by how efficiently it performs in real, integrated systems. This practical validation comes through pilot-scale installations and slipstream demonstrations, where a portion of actual flue gas is used to replicate industrial conditions. Such setups uncover the genuine behaviour of the material interface under realistic environments, offering critical insights into its performance [11]. In this context, it becomes essential to evaluate a wide range of materials, from advanced materials like MOFs and membranes to nanocomposites and other emerging options, to identify both their strengths and limitations when scaled up. Case study analyses highlight how efficiency, longevity, and cost-effectiveness are deeply interconnected, and how these outcomes depend strongly on the interplay between material properties, structural stability, and operating conditions. A deeper understanding of these parameters is crucial, as it enables us to design strategies that address engineering challenges and foster innovative solutions for successful full-scale deployment of carbon capture technologies [12].

6.4 Nanocomposite Sorbents: Leveraging Interfacial Synergy for Enhanced Performance

This class of sorbent materials largely includes amine-based systems, which are among the most mature solid materials developed for CO_2 capture. In these materials, amine compounds are dispersed on a high surface area porous support such as silica or carbon. They bring several advantages, the most important being the strong chemical affinity of amines for CO_2, combined with the structural and thermal stability provided by the inorganic framework. A key step in evaluating amine-functionalised materials is their pilot-scale testing under repeated TSA cycles, with much of this work carried out at the National Carbon Capture Centre (NCCC) [18]. In these tests, flue gas from coal-fired power plants is passed through a reactor packed with the nanocomposite material. To improve contact between the gas and the sorbent, the material is typically shaped into pellets or beads and loaded into a fluidised bed reactor. The critical interface in these systems exists between the amine molecules and the porous support surface. Achieving high efficiency requires careful tuning of amine loading at this interface. This is a delicate balance, ensuring uniform amine distribution inside the pores while minimising problems such as pore blockage or limited gas diffusion in bulk amine layers. At the same time, the support material itself must be robust, since good thermal and mechanical stability is essential for long-term performance [19].

The main challenge in scaling up lies in maintaining stable interactions at the interface, which often becomes highly vulnerable to degradation under real operating conditions. The most critical hurdles in practical demonstrations are linked to hydrothermal stress and oxidative degradation. In flue gas mixtures, the presence of steam, oxygen, and trace corrosive gases accelerates this problem. Amines, being prone to oxidation, form inactive species and gradually detach from the support during repeated adsorption–desorption cycles, leading to irreversible loss of performance. This highlights a key lesson: materials that look promising under ideal laboratory conditions often fail when exposed to the harsher realities of industrial environments. To overcome this, solutions must extend beyond simply re-engineering the interface [18]. A more holistic approach is needed, one that reinforces both the structure and stability of the system. For example, designing core-shell architectures can shield sensitive cores with protective, gas-permeable shells. Similarly, tethering strategies that chemically anchor amines to the support can prevent their detachment over time. Another promising avenue is polymeric encapsulation, which traps amines securely within the pores while still allowing CO_2 to pass through efficiently [20]. Together, these approaches can address the root causes of degradation and significantly enhance the long-term durability of the material.

Another major challenge revealed in this case study is the high-energy cost required for regenerating the sorbent material. The CO_2 capture process involves several steps, and many of them contribute to this drawback. While the strong bond between amines and the support enhances stability, it also demands significant energy to break the CO_2-amine bond during desorption. This extra energy load becomes a burden on the power plant's efficiency. Pelletisation, although useful for improving stability, adds another layer of complexity. The thermal conductivity of the pellets depends on how well the active material interacts with the binder [21]. However, when pellets are thicker, heat transfer becomes less efficient. As a result, regeneration requires either more time or higher temperatures, both of which raise energy consumption. Addressing this issue calls for the development of an optimised interfacial network, one that not only ensures mechanical stability but also supports effective heat transfer. This has inspired researchers to explore conductive binders and the incorporation of conductive nanoparticles, which bring multiple advantages: improved active pore accessibility, better mechanical strength, and enhanced thermal conductivity [18,22].

Overall, this case study highlights that the efficiency of a sorbent is determined not just by its chemical affinity for CO_2, but also by its thermophysical properties and structural design. The ultimate goal is to achieve a material with a "Goldilocks" binding energy, strong enough to capture CO_2 effectively, even from dilute flue gases, but weak enough to allow regeneration with minimal energy input.

6.5 MOFs: From Crystalline Powder to Engineered Contactors

This class of materials has emerged as a remarkable option for gas separation because of their high porosity and tunable structures. Because they are crystalline, their properties, such as pore size and active sites, can be precisely tailored in the laboratory to enhance selectivity and capacity for CO_2 capture. A notable example is Mg-MOF-74, a metal-organic framework with open metal sites that strongly adsorb CO_2 molecules. While highly efficient at the laboratory scale, this material faces serious challenges when translated to industrial applications, mainly due to complications at the metal-binder interface [23]. For large-scale use, especially in temperature/vacuum swing adsorption (TVSA) processes, the material must be shaped into durable pellets. Achieving this requires the use of suitable binders and carefully controlled compression methods, both of which introduce significant engineering hurdles [24].

During synthesis, this step often becomes the bottleneck of the entire process. While pelletising, the applied mechanical pressure can easily damage the delicate crystalline framework of MOFs, collapsing their pores and rendering them inactive for CO_2 capture. Even if the crystalline structure is preserved, another serious challenge arises from the non-porous nature of the binders. These binders can coat or block the pores, restricting gas diffusion and significantly reducing the accessible surface area of the MOF pellets. The result is a material with lower adsorption capacity and poor CO_2 capture efficiency. This persistent issue has been one of the key reasons why many promising MOF-based materials have failed to move beyond pilot-scale demonstrations [25].

To move forward, it is essential to redesign the engineering of MOFs with innovative approaches. One promising direction is the development of MOF-polymer composites, where MOFs are grown in situ within a polymer matrix. This strategy not only ensures better integration but also minimises the risk of pore blockage. Another solution lies in creating hierarchical structures with embedded macropores inside the pellets. These larger pores act as highways, allowing gases to bypass dense binder regions and quickly reach the micropores within the MOF crystals. Perhaps the most versatile approach is the use of advanced 3D printing technologies to directly shape MOFs into desired geometries, eliminating the need for binders. Such structures can maintain their own mechanical stability without external support. A further challenge is the moisture sensitivity of many MOFs [26]. This can be addressed either by designing hydrophobic frameworks or by tailoring pore chemistry so that the material repels water while still maintaining strong selectivity for CO_2. From these case studies, it becomes clear that the barriers to MOF performance and stability are not limited to engineering design alone; they also require chemically robust architectures. Achieving a balance between structural innovation and chemical resilience will be the key to successfully integrating MOFs into large-scale CCUS applications.

6.6 Advanced Membrane Technologies: Where the Interface is the Technology

The use of membranes for carbon capture is built on a simple yet powerful idea: creating a thin barrier that selectively lets CO_2 pass through while blocking all other gases. Unlike in other technologies where the interface plays only a supporting role, here it is the very heart of the process, the essence of how the membrane works [27]. A major focus in this field is the development of MMMs. The goal behind MMMs. is to combine the high selectivity of advanced filler materials with the cost-effectiveness and easy processability of polymers. By embedding these fillers uniformly within a polymer matrix, researchers aim to produce thin composite membranes that significantly outperform pristine polymers. Pilot-scale testing of MMMs, particularly in hollow-fibre module designs, is already underway at several facilities. These trials expose the membranes to flue gas slipstreams under real-world operating conditions, offering crucial insights into their efficiency, durability, and scalability for post-combustion carbon capture [28].

The performance efficiency of MMMs. depends almost entirely on the quality of the filler-polymer interface. Ideally, the fillers should be distributed uniformly within the polymer matrix, creating a continuous, defect-free membrane. In such a structure, the polymer chains adhere strongly to the surface of the fillers, ensuring selective and reliable CO_2 transport. However, achieving this level of precision in membrane design is extremely challenging. When the adhesion between filler and polymer is weak, defects such as voids or interfacial gaps often form between them. These imperfections compromise the selectivity of the membrane, as they allow not only CO_2, but also other gases, moisture, and impurities to pass through [29]. This morphological defect is commonly referred to as the "sieve-in-a-cage" phenomenon. Such problems have frequently been reported in MMMs: while they may show excellent results in controlled laboratory environments, their performance often deteriorates when scaled up to larger areas for industrial applications.

These operational setbacks compelled researchers to pursue more refined engineering strategies for designing efficient membranes. One of the first approaches was the surface functionalisation of nanofillers before incorporating them into the polymer matrix. In this method, fillers such as MOFs, zeolites, and others are chemically modified with functional groups, enabling them to form stronger bonds with the polymer. This minimises the likelihood of interfacial void formation and improves structural integrity. A second promising strategy involved the use of flexible and highly permeable polymers [26, 28]. Examples include polymers with intrinsic microporosity or those whose arrangement can be thermally rearranged. Such polymers demonstrate a strong ability to retain nanofillers while maintaining intimate contact, thereby enhancing overall membrane performance. While these solutions have shown promise, critical challenges still arise during practical applications. Among the most pressing issues are sorption and plasticisation. In real flue gas streams, the presence of water vapour often leads to its adsorption onto either the filler or the polymer, blocking diffusion pathways for CO_2.

Additionally, under high CO_2 concentrations, certain polymers undergo swelling. This increases polymer chain mobility, reducing membrane selectivity over prolonged operation. Addressing these long-term stability issues requires the development of chemically robust polymers with cross-linked architectures. Such matrices can resist both swelling and ageing, ensuring more reliable performance over time [26].

From this case study, it is clear that membrane technology for CCUS applications has advanced significantly, transforming from early-stage failures into systems capable of delivering synergistic performance. This progress highlights the relentless pursuit of solutions in the field, driven by both scientific innovation and practical necessity.

6.7 System Integration and Operational Realities

Translating advanced materials for carbon capture from the laboratory to real-world applications is far from straightforward. Even after achieving success in tailoring their intrinsic properties and shaping them into desired forms, integrating these materials into industrial systems demands navigating a complex and dynamic environment. A closer look at industrial operations makes it clear that having efficient material is essential, but it represents only one part of the equation. The ultimate success of a carbon capture system depends on a network of interdependent factors, operational cycles, physical constraints, and long-term durability, all working in synergy. Beyond the material itself, the way it interacts with the process plays a decisive role. Heat and mass transfer, energy requirements, and degradation mechanisms pose significant challenges for engineering design. Any weakness at this stage can lead to system failure, underscoring a critical lesson: it is not enough to evaluate a material in isolation. Instead, pilot-scale testing must examine the full process as a unified whole [30, 31].

This reality has shifted focus towards co-engineering, where materials and processes are designed in tandem to ensure efficiency, resilience, and durability. To fully understand this landscape, it is useful to break down the different aspects that govern system performance, one by one.

6.8 Material-Process Interface: Matching Properties to an Operational Cycle

Before moving towards large-scale operation, selecting the right material is a critical first step. The choice of material is tightly linked to the type of operational cycle, whether it is TSA, VSA, pressure swing adsorption (PSA), or any hybrid variant. Each cycle comes with its own unique requirements, meaning the material must be chosen not just for its intrinsic CO_2 capture ability, but also for its compatibility with the operating conditions. Take TSA as an example: here, regeneration is driven by heat. Therefore, the adsorbent must have high thermal conductivity to ensure uniform heat distribution across the reactor bed. This property is essential for reducing both the time and energy required for regeneration [32]. If a material is chosen solely for its high CO_2 uptake but has poor thermal conductivity, it will lead to the formation of cold spots in the bed, ultimately making regeneration inefficient. Over time, this inefficiency causes a gradual decline in performance.

In contrast, PSA is a pressure-driven process, where the speed of adsorption and desorption is paramount. Materials must allow CO_2 to diffuse rapidly into and out of their pores. Even if a sorbent shows excellent equilibrium capacity, if its kinetics are too slow, it cannot fully perform within the rapid cycles of PSA. In such cases, the material's potential remains underutilised, making it unfit for practical use in this process [33].

Another major challenge lies in the material-process interface, which often becomes a critical bottleneck for heat and mass transfer when moving from laboratory-scale studies to industrial applications. At the lab scale, achieving uniformity across small samples is relatively straightforward. However, scaling up introduces complexities; gradients of temperature, pressure, and concentration begin to appear and must be carefully managed. For instance, adsorption is an exothermic process, and the buildup of a temperature gradient severely deteriorates performance if not controlled. As the bed heats up, the affinity of the sorbent for CO_2 decreases due to the "heat of adsorption" effect, ultimately reducing its equilibrium capacity [34]. To counter this, industrial systems must incorporate advanced cooling mechanisms and structured packing designs to facilitate efficient heat transfer and maintain uniform operation.

Another complication is ensuring consistent gas flow throughout the adsorption bed. A common issue is the "channelling effect", where gases preferentially flow through low-resistance pathways, bypassing large sections of the material. This uneven flow significantly reduces overall efficiency. Reactor design, therefore, becomes a critical consideration. For example, fluidised beds provide excellent heat and mass transfer, but their effectiveness depends on materials with high attrition resistance to withstand constant particle collisions [35]. Addressing these intertwined challenges calls for a tandem approach, where materials and reactor designs are co-developed to complement one another, ensuring stable, efficient, and scalable performance.

6.9 Retrofitting vs New Builds: Designing for Existing Infrastructure

To meet the growing demand, carbon capture systems in the present and future will largely need to be retrofitted into existing industrial setups such as power plants, cement factories, and steel mills. However, putting this into practice comes with stringent constraints: spatial, operational, and financial, which make the selection of technology and system design a real challenge. Every industrial facility already has a channelised layout, with predefined flow rates, steam, and power systems. This means that when integrating a carbon capture "island" into such a setup, the installation must be done with minimal alterations and at the lowest possible cost. For example, deploying a TSA system requires the availability of a low-pressure steam supply. If the host plant does not already have this provision, auxiliary boilers would need to be installed. This not only consumes additional space, often scarce in existing plants, but also drives up overall costs. Such space limitations are one of the strongest arguments for the development of modular systems, which allow TSA and similar technologies to be installed more compactly and efficiently, making retrofitting a more practical option [36]. The regeneration of capture materials directly impacts the overall energy consumption of the system, as the added installations act as parasitic loads, creating a major concern for large-scale deployment. In fact, depending on the setup, conventional amine scrubbing alone can reduce a plant's net power output by 15–25%. This heavy energy demand during regeneration remains one of the most critical challenges for advanced material development. To address this, two main strategies are being explored: either designing materials with inherently lower regeneration energy requirements, making them easier to retrofit into existing plants, or reducing the heat of adsorption, which would allow regeneration to occur under near-ambient conditions [37].

Another pressing issue lies in the pressure drop across the capture unit, which often requires upgrading to more powerful and costly blowers. To minimise this, researchers are focusing on creating low-density packing forms of materials that enable smoother gas flow with minimal intrinsic resistance. Balancing these requirements highlights a key reality: retrofitting existing plants is far more challenging than designing entirely new facilities [38]. Yet, it is also the most urgent pathway, since large-scale carbon capture will primarily depend on integrating with the infrastructure we already have.

6.10 Long-Term Performance and Deactivation

After ensuring successful installation and operation, the final and perhaps most critical challenge lies in the long-term durability of the materials. Industrial units are designed to run continuously for months or even years, with only brief shutdowns for maintenance. For carbon capture materials, this means they must not only perform efficiently at the start but also maintain their performance over extended operational cycles under harsh chemical environments [39]. Material degradation, therefore, becomes an unavoidable concern. As discussed earlier, sorbents are especially susceptible to contaminants, which can irreversibly damage active sites. The major modes of degradation include oxidation, thermal breakdown, and physical ageing, the latter being particularly critical in membranes. In physical ageing, polymer chains gradually rearrange into denser configurations, which reduces permeability and, over time, undermines performance. To effectively address this issue, it is essential to develop a deep understanding of the underlying degradation mechanisms. Such insights not only allow researchers to predict the lifespan of materials but also guide the design of mitigation strategies, ensuring stable and reliable performance in real-world industrial operations [40].

To address the issue of long-term durability, it is essential to integrate material health monitoring and rejuvenation systems into existing plants. For instance, the deactivation of amine sorbents caused by corrosive flue gas contaminants can be partly mitigated through periodic washing to remove heat-stable salts. Similarly, materials that suffer from pore blockage due to the deep adsorption of impurities can be rejuvenated through regular burn-off cycles, which help clear the pores and restore performance. However, while these rejuvenation methods are effective, they also add complexity and downtime to the operation, which must be carefully weighed against both economic and operational considerations [22]. It is also important to recognise that every material has a finite lifetime, eventually requiring replacement. Therefore, the costs of replacement and disposal must be factored into the overall economic model of a carbon capture plant. This means performance assessments must extend beyond initial efficiency to also consider the half-life and durability of the material. Looking forward, research should focus on developing standardised tests under laboratory conditions that realistically simulate long-term industrial environments. Such protocols would allow more accurate predictions of material lifetime and better planning for maintenance schedules [22, 40].

In summary, a truly successful carbon capture system will combine slow material degradation, manageable maintenance and replacement strategies, and economic feasibility, ensuring both technical reliability and long-term sustainability for CCUS applications.

6.11 Overcoming Barriers and Charting the Future Course

The move from laboratory-scale research to full-scale industrial carbon capture is filled with technical and economic hurdles. Yet, ongoing efforts to uncover the root causes of these challenges and to create integrated solutions that combine scientific innovation with economic feasibility are steadily guiding progress toward meaningful success. Reaching the milestone of net-zero emissions demands careful navigation and well-judged trade-offs. This requires aligning system design with the realities of industrial-scale operations, managing the intricate interplay between system components, and advancing material interfaces to deliver dependable performance. In the end, progress will not rest on material development alone, but on a balanced portfolio that combines efficiency, cost-effectiveness, and long-term durability, while staying attuned to the evolving global ecosystem. At the centre of carbon capture lies the recognition of the "iron triangle" of cost, performance, and durability. In its early phases, most research emphasised improving material efficiency and selectivity in isolation, while long-term durability and economic viability were overlooked. This narrow lens led to the creation of materials that, despite excellent laboratory results, proved unrealistic for large-scale use due to cost constraints. Confronting this gap has strengthened the focus on the iron triangle. Achieving balance here requires a paradigm shift, where materials move from laboratory to industrial scale with careful focus on precursor choice (in terms of cost and safety), energy efficiency, scalability of synthesis, regeneration strategies and their energy needs, and overall cost-effectiveness.

Only through alignment of these dimensions can carbon capture technologies advance from promising trials to commercially viable solutions. This underscores the importance of rigorous techno-economic assessments to shape robust policy frameworks and stimulate public-private partnerships that can support the large-scale investments required. Such initiatives are the foundation for "first-of-a-kind" (FOAK) carbon capture projects, which are indispensable for moving technology beyond the research stage. Yet, high capital demands and risks tied to unproven systems often block commercialisation. Policies like the US 45Q tax credit provide essential support by offering direct payments per tonne of CO_2 captured and stored, making projects more financially viable. Likewise, government funding for pilot-scale demonstrations and related programmes plays a crucial role in strengthening investor trust by validating technologies under real-world operating conditions. To accelerate adoption, public and private sectors must come together, pooling expertise in technology, finance, and management, in order to distribute risks and advance progress more effectively. Beyond the steps being taken today, the long-term success of carbon capture technologies will hinge on breakthroughs in materials science, especially at the interface level. A powerful transformation is expected through the adoption of artificial intelligence (AI) and machine learning (ML) to scale-up operations with greater accuracy and efficiency. These computational approaches can anticipate results and identify obstacles before full-scale deployment, lowering risks and avoiding expensive failures. The integration of such algorithms will also reduce labour requirements while improving precision at every stage of processing. Another exciting frontier is the development of bio-inspired and self-healing materials, capable of repairing themselves and extending operational lifetimes. Alongside this, customised fabrication strategies are being designed for both point-source capture and direct air capture, given the unique property requirements of each. Together, these innovations suggest a future in which adaptive materials, intelligent design, and digital integration work in synergy, bringing carbon capture technologies closer to real-world, sustainable deployment.

6.12 Conclusion

The global effort to create effective carbon capture strategies has ignited a surge of innovation in advanced materials, driving a profound transformation across material science. This chapter has showcased the promising progress achieved in laboratories, while also revealing the obstacles and failures encountered when scaling these technologies for cost-effective industrial deployment. It is now clear that the greatest challenges, whether linked to time, reusability, or operational constraints, are not caused only by the intrinsic chemistry of the materials but are rooted in the broader management of the overall system. This makes engineering excellence at every stage indispensable, with the material-process interface emerging as the true foundation for practical applications. From nanoscale active sites to micro-scale integrations, every step forms a critical link in ensuring the success of carbon capture technologies.

Through case studies spanning nanocomposites, MOFs, and advanced membranes, it becomes evident that developing materials in isolation is no longer sufficient. Interfacial challenges, including efficiency, mechanical strength, and selectivity, are shaped not only by the inherent properties of the material but equally by careful engineering design. Alongside this, practical concerns such as synthesis routes, the use of binders, degradation from steam and contaminants, as well as issues of heat buildup and mass transfer, all reinforce the need for a holistic, system-level engineering framework. The central takeaway is that meaningful results can only be achieved when materials are co-optimised with the exact operating conditions of their target environment.

This chapter ultimately emphasises a shift from narrow material development towards a unified vision of carbon capture. Success cannot rely on laboratory achievements alone; it requires integrated systems where materials and processes complement one another, with interface optimisation placed at the forefront of research and design. Realising this goal demands a collaborative ecosystem of engineers, scientists, and economists. Their combined expertise must focus on navigating the "iron triangle" of cost, performance, and durability. By embracing this perspective, the field can bridge the so-called "valley of death," translating material breakthroughs into industrially viable solutions, while safeguarding both technological progress and the global climate.

REFERENCES

[1] K.W. Bandilla, Carbon capture and storage, in: Futur. Energy Improv. Sustain. Clean Options Our Planet, 2020: pp. 669–692. https://doi.org/10.1016/B978-0-08-102886-5.00031-1.

[2] Why carbon-capture tech is key to reaching climate goals | World Economic Forum. https://www.weforum.org/stories/2023/10/why-carbon-capture-is-key-to-reaching-climate-goals/.

[3] D.Y. Shu, S. Deutz, B.A. Winter, N. Baumgärtner, L. Leenders, A. Bardow, The role of carbon capture and storage to achieve net-zero energy systems: Trade-offs between economics and the environment, Renew. Sustain. Energy Rev. 178 (2023) 113246. https://doi.org/10.1016/J.RSER.2023.113246.

[4] M.E. Diego, M. Akram, J.M. Bellas, K.N. Finney, M. Pourkashanian, Making gas-CCS a commercial reality: The challenges of scaling up, Greenh. Gases Sci. Technol. 7 (2017) 778–801. https://doi.org/10.1002/GHG.1695.

[5] N. Wang, K. Akimoto, G.F. Nemet, What went wrong? Learning from three decades of carbon capture, utilization and sequestration (CCUS) pilot and demonstration projects, Energy Policy 158 (2021) 112546. https://doi.org/10.1016/J.ENPOL.2021.112546.

[6] R.L. Siegelman, P.J. Milner, E.J. Kim, S.C. Weston, J.R. Long, Challenges and opportunities for adsorption-based CO_2 capture from natural gas combined cycle emissions, Energy Environ. Sci. 12 (2019) 2161–2173. https://doi.org/10.1039/C9EE00505F.

[7] Advanced nanostructured materials for industrial carbon capture. https://www.renishaw.com/en/advanced-nanostructured-materials-for-industrial-carbon-capture—48202?srsltid=AfmBOort8lAEWu58jadjY4EOruXuS1ttDOk-1wCzkGbDOgelvVpE2MTk.

[8] H.H. Do, I. Rabani, H.B. Truong, Metal-organic framework-based nanomaterials for CO_2 storage: A review, Beilstein J. Nanotechnol. 1479 14 (2023) 964–970. https://doi.org/10.3762/BJNANO.14.79.

[9] N. Mittal, A. Samanta, P. Sarkar, R. Gupta, Postcombustion CO_2 capture using N-(3-trimethoxysilylpropyl) diethylenetriamine-grafted solid adsorbent, Energy Sci. Eng. 3 (2015) 207–220. https://doi.org/10.1002/ESE3.64.

[10] W. Chaikittisilp, H.J. Kim, C.W. Jones, Mesoporous Alumina-Supported Amines as Potential Steam-Stable Adsorbents for Capturing CO_2 from Simulated Flue Gas and Ambient Air, Energy and Fuels 25 (2011) 5528–5537. https://doi.org/10.1021/EF201224V.

[11] C. Charalambous, E. Moubarak, J. Schilling, E. Sanchez Fernandez, J.Y. Wang, L. Herraiz, F. Mcilwaine, S.B. Peh, M. Garvin, K.M. Jablonka, S.M. Moosavi, J. Van Herck, A.Y. Ozturk, A. Pourghaderi, A.Y. Song, G. Mouchaham, C. Serre, J.A. Reimer, A. Bardow, B. Smit, S. Garcia, A holistic platform for accelerating sorbent-based carbon capture, Nature 632 (2024) 89–94. https://doi.org/10.1038/S41586-024-07683-8;TECHMETA.

[12] D. DeSantis, J.A. Mason, B.D. James, C. Houchins, J.R. Long, M. Veenstra, Techno-economic Analysis of Metal–Organic Frameworks for Hydrogen and Natural Gas Storage, Energy and Fuels 31 (2017) 2024–2032. https://doi.org/10.1021/ACS.ENERGYFUELS.6B02510.

[13] M. Wang, H. Guo, C. Wang, L. Xiong, X. Chen, F. Peng, S. Yao, H. Li, H. Zhang, X. Chen, Scale-up synthesis and shaping of MIL-160(Al) applicable for efficient CO_2 adsorption and separation, Sep. Purif. Technol. 360 (2025) 131091. https://doi.org/10.1016/J.SEPPUR.2024.131091.

[14] B. Singh, M. Kemell, T. Repo, Porous carbon pellets for physical adsorption of CO_2: size and shape effect, Mater. Adv. 5 (2024) 7601–7608. https://doi.org/10.1039/d4ma00703d.

[15] D. Andirova, Y. Lei, X. Zhao, S. Choi, Functionalization of Metal-Organic Frameworks for Enhanced Stability under Humid Carbon Dioxide Capture Conditions, ChemSusChem 8 (2015) 3405–3409. https://doi.org/10.1002/CSSC.201500580.

[16] J. Hu, X. Gu, L.C. Lin, B.R. Bakshi, Toward Sustainable Metal-Organic Frameworks for Post-Combustion Carbon Capture by Life Cycle Assessment and Molecular Simulation, ACS Sustain. Chem. Eng. 9 (2021) 12132–12141. https://doi.org/10.1021/ACSSUSCHEMENG.1C03473.

[17] P. Wang, Z. Pan, Z. Liu, L. Shang, M.R. Soltanian, Z. Zhang, Life cycle assessment of CO_2 capture through pressure swing adsorption using MOF-74, Gas Sci. Eng. 133 (2025) 205497. https://doi.org/10.1016/J.JGSCE.2024.205497.

[18] H.J. Moon, J.M.Y. Carrillo, C.W. Jones, Distribution and Mobility of Amines Confined in Porous Silica Supports Assessed via Neutron Scattering, NMR, and MD Simulations: Impacts on CO_2 Sorption Kinetics and Capacities, Acc. Chem. Res. 56 (2023) 2620–2630. https://doi.org/10.1021/ACS.ACCOUNTS.3C00363.

[19] G. Zhang, P. Zhao, Y. Xu, Z. Yang, H. Cheng, Y. Zhang, Structure Property-CO_2 Capture Performance Relations of Amine-Functionalized Porous Silica Composite Adsorbents, ACS Appl. Mater. Interfaces 10 (2018) 34340–34354. https://doi.org/10.1021/ACSAMI.8B13069.

[20] S. Gadipelli, H.A. Patel, Z. Guo, S. Gadipelli, H.A. Patel, Z. Guo, An Ultrahigh Pore Volume Drives Up the Amine Stability and Cyclic CO_2 Capacity of a Solid-Amine@Carbon Sorbent, Adv. Mater. 27 (2015) 4903–4909. https://doi.org/10.1002/ADMA.201502047.

[21] R. Veneman, N. Frigka, W. Zhao, Z. Li, S. Kersten, W. Brilman, Adsorption of H_2O and CO_2 on supported amine sorbents, Int. J. Greenh. Gas Control 41 (2015) 268–275. https://doi.org/10.1016/J.IJGGC.2015.07.014.

[22] M.J. Bos, V. Kroeze, S. Sutanto, D.W.F. Brilman, Evaluating Regeneration Options of Solid Amine Sorbent for CO_2 Removal, Ind. Eng. Chem. Res. 57 (2018) 11141–11153. https://doi.org/10.1021/ACS.IECR.8B00768.

[23] B. Yeskendir, J.P. Dacquin, Y. Lorgouilloux, C. Courtois, S. Royer, J. Dhainaut, From metal–organic framework powders to shaped solids: recent developments and challenges, Mater. Adv. 2 (2021) 7139–7186. https://doi.org/10.1039/D1MA00630D.

[24] T.C. Wang, A.M. Wright, W.J. Hoover, K.J. Stoffel, R.K. Richardson, S. Rodriguez, R.C. Flores, J.P. Siegfried, N.A. Vermeulen, P.E. Fuller, M.H. Weston, O.K. Farha, W. Morris, Surviving Under Pressure: The Role of Solvent, Crystal Size, and Morphology During Pelletization of Metal-Organic Frameworks, ACS Appl. Mater. Interfaces 13 (2021) 52106–52112. https://doi.org/10.1021/ACSAMI.1C09619.

[25] E. Tsalaporta, J.M.D. MacElroy, A comparative study of the physical and chemical properties of pelletized HKUST-1, ZIF-8, ZIF-67 and UiO-66 powders, Heliyon 6 (2020) e04883. https://doi.org/10.1016/J.HELIYON.2020.E04883.

[26] S. Lawson, M. Snarzyk, D. Hanify, A.A. Rownaghi, F. Rezaei, Development of 3D-Printed Polymer-MOF Monoliths for CO_2 Adsorption, Ind. Eng. Chem. Res. 59 (2020) 7151–7160. https://doi.org/10.1021/acs.iecr.9b05445.

[27] M. Sandru, T.J. Kim, W. Capala, M. Huijbers, M.B. Hägg, Pilot scale testing of polymeric membranes for CO_2 capture from coal fired power plants, Energy Procedia 37 (2013) 6473–6480. https://doi.org/10.1016/j.egypro.2013.06.577.

[28] Y. Liu, Y. Sun, F. Sun, H. Li, J. Guan, L. Bai, Z. Gao, T. Li, F. Fan, G. He, C. Ma, High-performance ester-crosslinked polymers of intrinsic microporosity membranes with enhanced CO_2 selectivity and plasticization resistance, Sep. Purif. Technol. 364 (2025) 132447. https://doi.org/10.1016/J.SEPPUR.2025.132447.

[29] G. Kupgan, A.G. Demidov, C.M. Colina, Plasticization behavior in polymers of intrinsic microporosity (PIM-1): A simulation study from combined Monte Carlo and molecular dynamics, J. Memb. Sci. 565 (2018) 95–103. https://doi.org/10.1016/J.MEMSCI.2018.08.004.

[30] S.H.A.M. Leenders, G. Pankratova, J. Wijenberg, J. Romanuka, F. Gharavi, J. Tsou, M. Infantino, L. van Haandel, S. van Paasen, P.E. Just, Amine Adsorbents Stability for Post-Combustion CO_2 Capture: Determination and Validation of Laboratory Degradation Rates in a Multi-staged Fluidized Bed Pilot Plant, ChemSusChem 16 (2023) e202300930. https://doi.org/10.1002/CSSC.202300930.

[31] A. Tatarczuk, M. Tańczyk, L. Więcław-Solny, J. Zdeb, Pilot plant results of amine-based carbon capture with heat integrated stripper, Appl. Energy 367 (2024) 123416. https://doi.org/10.1016/J.APENERGY.2024.123416.

[32] J.Y. Lai, L.H. Ngu, S.S. Hashim, A review of CO_2 adsorbents performance for different carbon capture technology processes conditions, Greenh. Gases Sci. Technol. 11 (2021) 1076–1117. https://doi.org/10.1002/GHG.2112.

[33] M.G. Plaza, F. Rubiera, C. Pevida, Evaluating the Feasibility of a TSA Process Based on Steam Stripping in Combination with Structured Carbon Adsorbents To Capture CO_2 from a Coal Power Plant, Energy and Fuels 31 (2017) 9760–9775. https://doi.org/10.1021/ACS.ENERGYFUELS.7B01508.

[34] F. Rezaei, M. Grahn, Thermal Management of Structured Adsorbents in CO_2 Capture Processes, Ind. Eng. Chem. Res. 51 (2012) 4025–4034. https://doi.org/10.1021/IE201057P.

[35] C. Dhoke, A. Zaabout, S. Cloete, S. Amini, Review on Reactor Configurations for Adsorption-Based CO_2 Capture, Ind. Eng. Chem. Res. 60 (2021) 3779–3798. https://doi.org/10.1021/ACS.IECR.0C04547.

[36] D. Obi, S. Onyekuru, A. Orga, Minimizing Carbon Capture Costs in Power Plants: A Novel Dimensional Analysis Framework for Techno-Economic Evaluation of Oxyfuel Combustion, Pre-combustion, and Post-combustion Capture Systems, Energy Sci. Eng. 13 (2025) 1749–1770. https://doi.org/10.1002/ESE3.2089.

[37] M.H. Islam, S.R. Patlolla, A Review on the State-of-the-Art and Commercial Status of Carbon Capture Technologies, Energies 2025, Vol. 18, Page 3937 18 (2025) 3937. https://doi.org/10.3390/EN18153937.

[38] O.H.P. Gunawardene, C.A. Gunathilake, K. Vikrant, S.M. Amaraweera, Carbon Dioxide Capture through Physical and Chemical Adsorption Using Porous Carbon Materials: A Review, Atmos. 2022, Vol. 13, Page 397 13 (2022) 397. https://doi.org/10.3390/ATMOS13030397.

[39] C. Rosu, S.H. Pang, A.R. Sujan, M.A. Sakwa-Novak, E.W. Ping, C.W. Jones, Effect of Extended Aging and Oxidation on Linear Poly(propylenimine)-Mesoporous Silica Composites for CO_2 Capture from Simulated Air and Flue Gas Streams, ACS Appl. Mater. Interfaces 12 (2020) 38085–38097. https://doi.org/10.1021/ACSAMI.0C09554.

[40] N. Nityashree, G. V. Manohara, M.M. Maroto-Valer, S. Garcia, Advanced High-Temperature CO_2 Sorbents with Improved Long-Term Cycling Stability, ACS Appl. Mater. Interfaces 12 (2020) 33765–33774. https://doi.org/10.1021/ACSAMI.0C08652.

Chapter 7

FUTURE PROSPECTS OF CARBON CAPTURE, UTILISATION, AND STORAGE

Abstract: Carbon Capture, Utilisation, and Storage (CCUS) is increasingly recognised as a critical component of global strategies aimed at achieving net-zero and net-negative greenhouse gas emissions. While renewable energy deployment and energy-efficiency improvements are essential, they are insufficient to fully decarbonise hard-to-abate sectors such as cement, steel, chemicals, and long-distance transport. CCUS uniquely addresses this gap by enabling direct management of CO_2 emissions at point sources and from the atmosphere, while also supporting carbon removal pathways. This chapter presents a comprehensive and forward-looking assessment of the prospects of CCUS, emphasising scientific, technological, and systemic developments required for large-scale deployment. Particular attention is given to advances in carbon capture technologies, where next-generation materials and interface engineering are expected to reduce energy penalties and improve operational durability significantly. Emerging approaches such as electrochemically mediated capture, hybrid capture systems, and direct air capture are discussed in the context of future energy systems dominated by renewable electricity. The chapter further examines the evolving role of carbon utilisation in enabling circular carbon economies, highlighting progress in CO_2-to-fuels, chemicals, mineralisation, and hybrid bio-electrochemical pathways. Carbon storage is addressed as the backbone of CCUS, with emphasis on geological capacity, long-term integrity, monitoring, and public trust. The integration of CCUS with energy and industrial systems, including hydrogen production and industrial clusters, is identified as a key determinant of economic viability. Finally, the chapter discusses policy, economic, and societal dimensions shaping CCUS deployment, as well as the critical role of CCUS-enabled negative-emission technologies in long-term climate stabilisation. Overall, the chapter positions CCUS as a scientifically robust and strategically indispensable element of future low-carbon and climate-resilient energy systems.

Keywords: Carbon Capture, Utilisation and Storage, Negative Emissions, Energy System Integration, Climate Mitigation Strategies, Future and Challenges

7.1 Introduction

The accelerating pace of global climate change, driven predominantly by anthropogenic CO_2 emissions, has imposed unprecedented demands on the scientific, technological, and policy communities to develop mitigation strategies that are not only effective but also scalable, durable, and economically viable. Despite substantial progress in renewable energy deployment, electrification, and energy-efficiency improvements, global CO_2 emissions remain at historically elevated levels. This apparent paradox underscores a fundamental reality: modern energy and industrial systems are structurally complex and deeply embedded in carbon-intensive processes that cannot be fully decarbonised through renewable substitution or electrification alone. A significant fraction of global emissions originates from so-called hard-to-abate sectors, including cement, iron and steel production, chemicals manufacturing, petroleum refining, and long-distance transportation. In these sectors, CO_2 emissions are not merely the by-product of energy use but are intrinsically linked to process chemistry, thermodynamic constraints, and material transformation pathways. For example, calcination reactions in cement manufacturing and reduction reactions in metallurgical processes inherently generate CO_2, irrespective of the energy source employed. Consequently, mitigation strategies that rely exclusively on clean energy deployment are insufficient to achieve deep decarbonisation across the full industrial landscape.

Within this context, CCUS has emerged as a critical enabling technology for managing carbon emissions at scale. Unlike mitigation approaches that focus solely on reducing future emissions, CCUS directly addresses CO_2 that is already produced, either at concentrated point sources or from dilute streams such as ambient air. This capability positions CCUS uniquely within the portfolio of climate mitigation technologies, as it can be applied to existing infrastructure while simultaneously supporting the transition to low-carbon systems. The relevance of CCUS extends beyond emission avoidance to encompass carbon removal. Technologies such as bioenergy with carbon capture and storage (BECCS) and direct air capture with storage (DACCS) enable net-negative emissions by actively removing CO_2 from the atmosphere. These approaches are increasingly embedded in long-term climate stabilisation pathways, particularly in scenarios aiming to limit global temperature rise to well below 2°C, and preferably to 1.5°C. Integrated assessment models consistently demonstrate that achieving these targets without large-scale deployment of CCUS is either prohibitively expensive or technically infeasible. Many such model projects' cumulative CO_2 storage requirements are on the order of hundreds of gigatonnes by the end of the 21st century, underscoring the scale at which CCUS must operate. These projections highlight CCUS not as a marginal or transitional option, but as a core pillar of global climate strategy.

Nevertheless, the technology has historically faced substantial criticism. High capital and operating costs, significant energy penalties associated with capture and compression, uncertainties related to long-term storage integrity, and concerns regarding technological maturity have all constrained widespread deployment. In addition, CCUS has often been perceived as a means of prolonging fossil fuel dependence rather than facilitating systemic transformation, further complicating public acceptance.

In response to these challenges, the scientific and engineering communities have undergone a notable shift in research focus. Rather than pursuing incremental optimisation of legacy technologies—such as conventional solvent-based absorption systems—efforts are increasingly directed toward transformative innovation rooted in materials science, interface engineering, and system-level integration. This shift reflects a growing recognition that performance bottlenecks in CCUS are governed not solely by bulk material properties or process configurations, but by phenomena occurring at interfaces: gas-solid adsorption surfaces, liquid–gas absorption boundaries, electrochemical interfaces, and rock–fluid interactions in geological storage formations. Advances in advanced porous materials, electrochemically active sorbents, hybrid capture systems, and interface-tailored catalysts offer pathways to reduce capture energy demand substantially, improve selectivity under realistic operating conditions, and enhance material durability over extended cycling. Similarly, innovations in utilisation technologies increasingly focus on controlling interfacial reaction environments to improve conversion efficiency and product selectivity. In the context of storage, improved understanding of geochemical and geomechanical interfaces is critical for ensuring long-term containment and environmental safety. Equally important is the evolving conceptualisation of CCUS itself. It is no longer viewed as a standalone, end-of-pipe solution applied in isolation to individual facilities. Instead, CCUS is increasingly understood as a system-level technology that must be integrated within broader energy and industrial ecosystems. Emerging deployment models emphasise industrial clusters and shared infrastructure, where multiple emitters connect to centralised capture, transport, utilisation, and storage networks. Such configurations reduce costs through economies of scale, facilitate knowledge transfer, and enable coordinated risk management. This evolution also reflects a broader paradigm shift in how CO_2 is perceived. Rather than being treated exclusively as a waste product to be sequestered, CO_2 is increasingly regarded as a managed carbon resource. In this view, carbon flows can be deliberately directed towards recycling, conversion, or permanent immobilisation, depending on system-level optimisation criteria. This perspective underpins the development of circular carbon economies, in which CCUS plays a central role in closing material loops and reducing reliance on virgin fossil carbon. At the same time, the future of CCUS cannot be decoupled from questions of governance, policy design, and societal trust. Long-term deployment requires robust regulatory frameworks, transparent monitoring and verification mechanisms, and clear liability structures. Public acceptance depends on credible evidence of safety, environmental integrity, and alignment with broader sustainability objectives. Consequently, scientific and technological progress must be accompanied by institutional and social innovation.

This chapter examines the prospects of CCUS through a comprehensive and interdisciplinary lens. Emphasis is placed on emerging scientific trends, particularly those related to advanced materials and interfacial phenomena that underpin performance improvements across the CCUS value chain. Capture, utilisation, and storage are analysed as interdependent components of an integrated system rather than as isolated technologies. Their future evolution is discussed in the context of global energy transitions, industrial decarbonisation strategies, and long-term climate governance. Through this perspective, CCUS is positioned not as a temporary compromise but as a scientifically grounded and strategically indispensable element of sustainable climate mitigation pathways.

7.2 Future Directions in Carbon Capture Technologies

Carbon capture remains the most technologically intensive and economically dominant component of the CCUS value chain, often accounting for more than 60–70% of the total system cost and energy consumption in current implementations. As a result, the long-term feasibility and scalability of CCUS depend critically on advances in capture technologies that can simultaneously reduce energy penalties, enhance operational robustness, and maintain performance under realistic industrial conditions. Future progress in this domain will be driven less by incremental improvements to established processes and more by fundamental shifts in materials design, process integration, and system control. A defining trend in next-generation capture technologies is the transition from bulk material optimisation to interface-centric design. In solid sorbent-based systems, CO_2 capture is governed by molecular interactions at gas-solid interfaces, where surface functional groups, pore topology, and local electrostatic environments collectively determine adsorption thermodynamics and kinetics. Future materials will increasingly exploit hierarchical pore architectures, spatially graded surface chemistries, and engineered heterointerfaces to achieve high selectivity in the presence of moisture and competing gas species. Consequently, research priorities are shifting away from maximising absolute adsorption capacity towards optimising working capacity, cyclic stability, and low-energy regeneration under continuous operation.

Advanced porous materials, including MOFs, COFs, and engineered carbonaceous solids, are expected to remain central to future capture strategies. However, their successful translation beyond laboratory-scale demonstrations depends on addressing persistent challenges such as hydrothermal instability, framework degradation, and synthesis scalability. Interface engineering approaches, such as composite formation, defect modulation, polymer-framework hybrids, and surface passivation, are anticipated to play a crucial role in enhancing material durability while preserving adsorption performance. Electrochemically mediated capture represents a particularly transformative direction for carbon capture technology. By replacing thermally driven regeneration with electrically controlled adsorption–desorption cycles, electro-swing systems offer the potential to decouple capture energy demand from heat sources and align it directly with low-carbon electricity supply. This capability is especially attractive in energy systems characterised by high penetration of intermittent renewable power. Future development will depend on the design of redox-active materials with stable electrochemical interfaces, minimal side reactions, and long cycle lifetimes, as well as on system architectures that enable efficient ion and electron transport. DAC is expected to assume increasing importance in future climate mitigation pathways, particularly for addressing diffuse emissions and historical CO_2 accumulation. Although DAC is inherently energy-intensive due to the low concentration of CO_2 in ambient air, advances in ultra-selective sorbents, modular reactor configurations, and integration with low-grade or waste heat sources are anticipated to reduce costs progressively. Large-scale deployment will further benefit from standardisation, mass manufacturing, and learning-by-doing effects.

Finally, hybrid capture systems that integrate membranes, adsorption, and absorption processes are likely to become more prevalent. Such systems enable process intensification, reduced equipment footprint, and improved overall energy efficiency. The growing role of digitalisation through machine learning-assisted material discovery, process optimisation, and real-time system control will further accelerate innovation and facilitate the transition of advanced capture technologies from pilot-scale demonstrations to commercial deployment.

7.3 Carbon Utilisation: Expanding the Scope of Circular Carbon Systems

Carbon utilisation has the potential to fundamentally reshape the economic and societal perception of CCUS by transforming CO_2 from a waste product into a feedstock for value-added applications. While utilisation alone cannot absorb the total volume of captured CO_2 required for climate mitigation, its future role as a complementary pathway is increasingly recognised. One of the most promising utilisation routes involves the conversion of CO_2 into fuels and chemicals through thermochemical, electrochemical, or hybrid processes. When combined with low-carbon hydrogen, captured CO_2 can be converted into methanol, synthetic hydrocarbons, and oxygenated chemicals that serve as building blocks for the chemical industry. These pathways enable partial closure of the carbon loop, reducing dependence on fossil-derived feedstocks. Advances in catalysis and reaction engineering will govern the future of CO_2-to-chemicals technologies. Interfacial phenomena at catalyst surfaces determine reaction selectivity, energy efficiency, and catalyst lifetime. Future research will increasingly focus on stabilising active sites, controlling reaction microenvironments, and suppressing competing pathways. Reactor designs that enable efficient heat and mass transfer will be critical for scaling these processes. Electrochemical CO_2 reduction is particularly attractive due to its modularity and compatibility with renewable electricity. However, challenges related to electrode degradation, product separation, and current density limitations remain substantial. Interface engineering at the electrode-electrolyte boundary will play a decisive role in overcoming these barriers.

Mineralisation-based utilisation offers an alternative pathway with permanent storage benefits. Carbonated construction materials, aggregates, and supplementary cementitious materials can incorporate significant quantities of CO_2. Future development will focus on accelerating mineralisation kinetics, integrating CO_2 curing into industrial workflows, and establishing standardised performance metrics. Biological and hybrid utilisation routes may expand in niche applications, particularly as advances in synthetic biology improve productivity and selectivity. While scalability remains a challenge, hybrid bio-electrochemical systems represent a promising frontier for future research.

7.4 Integration of CCUS with Energy and Industrial Systems

The future effectiveness of CCUS depends fundamentally on its integration within broader energy and industrial systems, rather than its deployment as an isolated technological add-on. As energy systems transition towards low-carbon configurations, CCUS must be embedded in system-level architectures that optimise energy use, material flows, and infrastructure investments. This integrated perspective is essential for reducing costs, enhancing operational flexibility, and maximising climate mitigation impact. In the energy sector, CCUS plays a critical role in enabling system reliability during the transition to high shares of variable renewable energy. Power generation equipped with carbon capture can provide dispatchable, low-carbon electricity, complementing intermittent sources such as wind and solar. This firm capacity is particularly valuable for maintaining grid stability, reducing reliance on unabated fossil fuel peaker plants, and minimising curtailment of renewable generation. Furthermore, CCUS-equipped facilities can be operated flexibly, adjusting capture rates in response to electricity market conditions and grid demand, thereby improving overall system efficiency.

A particularly important dimension of integration lies in the coupling of CCUS with hydrogen production. Hydrogen is widely regarded as a key energy carrier for decarbonising industry, transport, and energy storage. In the near to medium term, hydrogen produced from fossil fuels with carbon capture—often referred to as blue hydrogen—offers a scalable pathway to establish hydrogen supply chains while renewable-based green hydrogen capacity expands. Integrating CCUS into hydrogen production systems allows for substantial emission reductions while leveraging existing infrastructure and expertise. Over time, hybrid systems combining blue and green hydrogen can support a gradual and economically viable transition. In industrial contexts, CCUS integration is most effective when implemented through cluster-based deployment models. Industrial clusters bring together multiple emission sources, such as cement plants, refineries, steel mills, and chemical facilities, within a shared geographic region. By utilising common CO_2 capture, transport, and storage infrastructure, these clusters reduce unit costs, distribute financial and operational risks, and enable coordinated planning. Such integration also facilitates industrial symbiosis, where captured CO_2 from one process can serve as an input for utilisation pathways in another. Digitalisation and system-level optimisation further enhance integration. Advanced process control, real-time monitoring, and digital twins enable coordinated operation across capture units, energy systems, and downstream utilisation or storage facilities. These tools allow operators to respond dynamically to changes in energy prices, demand patterns, and operational constraints.

Overall, integrating CCUS with energy and industrial systems transforms it from a standalone mitigation technology into a strategic enabler of low-carbon transitions. By aligning CCUS deployment with energy system evolution, industrial restructuring, and infrastructure planning, its role becomes not only technically effective but also economically and systemically indispensable.

7.5 Policy, Economics, and Societal Dimensions

The large-scale deployment of CCUS is governed not only by technological readiness but also by the policy, economic, and societal environments in which the technology operates. Even as scientific and engineering advances continue to reduce technical barriers, the absence of coherent policy frameworks and durable economic incentives remains one of the principal constraints on CCUS implementation. Future CCUS deployment will therefore depend critically on the alignment of regulatory instruments, market mechanisms, and public engagement strategies. From a policy perspective, carbon pricing mechanisms, such as carbon taxes and emissions trading systems, are widely recognised as foundational tools for creating long-term demand for CCUS. By internalising the social cost of carbon, these instruments improve the relative competitiveness of capture technologies and provide investment signals for infrastructure development. However, carbon prices alone are often insufficient to stimulate early-stage deployment, particularly in capital-intensive sectors. As a result, complementary policy measures, including tax credits, direct subsidies, and contracts for difference, will continue to play a central role. Contracts for difference, in particular, can provide revenue certainty by compensating the cost differential between low-carbon and conventional production pathways, thereby reducing investment risk. Economic viability will also be shaped by cost reductions achieved through technological learning and system integration. Historical experience with energy technologies demonstrates that costs decline as cumulative deployment increases, driven by learning curves, process optimisation, and supply-chain maturation. For CCUS, additional cost reductions are expected from standardisation of capture units, modular system designs, and shared transport and storage infrastructure. Industrial clustering models, where multiple emitters access common pipelines and storage sites, offer particularly strong potential for economies of scale and risk sharing.

Beyond policy and economics, societal acceptance represents a decisive factor in the long-term success of CCUS. Public concerns regarding safety, environmental integrity, and the perceived role of CCUS in extending fossil fuel use have influenced project approvals and regulatory decisions in several regions. Addressing these concerns requires transparent governance structures, rigorous monitoring and verification systems, and clear communication of risks and benefits. Demonstrating the safety and permanence of geological storage, in particular, is essential for building trust. Public engagement must extend beyond information dissemination to involve meaningful stakeholder participation in project planning and decision-making. When CCUS is framed as part of a broader sustainability strategy, supporting industrial decarbonisation, job preservation, and regional economic development, public acceptance tends to increase. Ultimately, the future of CCUS will depend on the successful integration of technical innovation with credible policy design, sound economic incentives, and sustained societal trust.

7.6 CCUS and Negative Emissions

Negative-emission technologies (NETs) are increasingly recognised as indispensable components of long-term climate stabilisation strategies, particularly in scenarios aimed at limiting global temperature rise to well below 2°C and achieving net-zero or net-negative emissions. Within this context, CCUS provides the technological foundation for two of the most prominent NETs: BECCS and direct air capture with storage (DACCS). These approaches extend the role of CCUS beyond emission avoidance to active removal of CO_2 from the atmosphere, addressing both residual emissions and historical carbon accumulation. BECCS combines biomass-based energy production with CO_2 capture and geological storage, thereby creating a net transfer of carbon from the atmosphere to long-term reservoirs. The future potential of BECCS lies in its ability to deliver negative emissions while simultaneously producing electricity, heat, or fuels. However, large-scale deployment raises significant sustainability concerns. Land-use competition with food production, impacts on biodiversity, water demand, and fertiliser use represent critical constraints that must be carefully managed. The climate benefits of BECCS are highly sensitive to biomass sourcing, cultivation practices, and supply-chain emissions, making a comprehensive life-cycle assessment essential for evaluating net mitigation outcomes. Direct air capture with storage offers a fundamentally different pathway to negative emissions by removing CO_2 directly from ambient air, independent of emission source location. This characteristic provides exceptional flexibility in siting and scalability, allowing DACCS systems to be deployed wherever low-carbon energy and suitable storage sites are available.

Nevertheless, DACCS faces inherent thermodynamic and economic challenges due to the low concentration of atmospheric CO_2. Future feasibility will depend on continued advances in ultra-selective capture materials, low-energy regeneration processes, and integration with renewable or waste energy sources. As with BECCS, life-cycle emissions associated with energy supply and material production must be rigorously accounted for to ensure genuine net removals.

Across both BECCS and DACCS, responsible deployment of negative-emission technologies requires robust sustainability frameworks. Life-cycle analysis must extend beyond greenhouse gas balances to include land-use change, ecosystem impacts, water consumption, and socio-economic effects. Governance mechanisms, monitoring, and verification systems, and long-term liability arrangements are also essential to ensure environmental integrity and public trust. Ultimately, CCUS-enabled negative emissions should be viewed as a complement to, rather than a substitute for, aggressive emission reductions, serving as a critical tool for balancing residual emissions in a climate-constrained future.

ACKNOWLEDGEMENTS

Dr. Irfan sincerely expresses gratitude to all those who supported and guided me throughout this work. I am deeply thankful to my parents and siblings for their constant encouragement, patience, and unwavering support. I owe special thanks to my Ph.D. supervisor, Dr. Vijay Kumar, for his invaluable guidance, motivation, and insightful discussions, which greatly shaped my scientific understanding and research outlook. I gratefully acknowledge the academic and research support provided by the National Institute of Technology Srinagar and the University of the Free State. I also thank Indian Institute of Technology Ropar and ARES Systems for providing a stimulating research environment and valuable exposure to hydrogen generation technologies and translational research. I further acknowledge Indian Institute of Technology Bombay for access to advanced electron microscopy facilities and an inspiring research atmosphere that contributed significantly to this monograph. I sincerely thank the Council of Scientific and Industrial Research (CSIR), Government of India, for financial support through the Research Associate Fellowship at the Islamic University of Science and Technology. Finally, I extend my heartfelt thanks to my collaborators, colleagues, and students for their valuable discussions, technical support, and encouragement throughout this work.

The authors sincerely acknowledge all the scientists, researchers, and contributors whose work has directly/indirectly supported the development of this book. Most of the figures and graphical illustrations were prepared by the authors using CorelDRAW, Photoshop, and various AI-assisted tools. Some images and scientific contents adapted from research articles and authentic sources have been properly cited and acknowledged wherever applicable. The authors are grateful to all stakeholders for their valuable contributions to the scientific community. This book has been prepared solely for educational and academic purposes and not for commercial profit. The work was supported under the SPARC Project (P3808) led by Prof. Santosh K. Tiwari.

www.ingramcontent.com/pod-product-compliance
Lightning Source LLC
LaVergne TN
LVHW080526210926
848481LV00022B/3324